T0155647

# Automated Market Makers

## A Practical Guide to Decentralized Exchanges and Cryptocurrency Trading

Miguel Ottina
Peter Johannes Steffensen
Jesper Kristensen

Apress®

*Automated Market Makers: A Practical Guide to Decentralized Exchanges and Cryptocurrency Trading*

Miguel Ottina
Mendoza, Argentina

Jesper Kristensen
New York, NY, USA

Peter Johannes Steffensen
Aarhus, Denmark

ISBN-13 (pbk): 978-1-4842-8615-9
https://doi.org/10.1007/978-1-4842-8616-6

ISBN-13 (electronic): 978-1-4842-8616-6

Managing Director, Apress Media LLC: Welmoed Spahr
Acquisitions Editor: James Robinson-Prior
Development Editor: James Markham
Coordinating Editor: Gryffin Winkler

Cover image designed by eStudioCalamar (www.estudiocalamar.com)

Distributed to the book trade worldwide by Springer Science+Business Media New York, 233 Spring Street, 6th Floor, New York, NY 10013. Phone 1-800-SPRINGER, fax (201) 348-4505, e-mail orders-ny@springer-sbm.com, or visit www.springeronline.com. Apress Media, LLC is a California LLC and the sole member (owner) is Springer Science + Business Media Finance Inc (SSBM Finance Inc). SSBM Finance Inc is a **Delaware** corporation.

For information on translations, please e-mail booktranslations@springernature.com; for reprint, paperback, or audio rights, please e-mail bookpermissions@springernature.com.

Apress titles may be purchased in bulk for academic, corporate, or promotional use. eBook versions and licenses are also available for most titles. For more information, reference our Print and eBook Bulk Sales web page at http://www.apress.com/bulk-sales.

Any source code or other supplementary material referenced by the author in this book is available to readers on GitHub. For more detailed information, please visit http://www.apress.com/source-code.

Printed on acid-free paper

# Table of Contents

# About the Authors

**Miguel Ottina** is a mathematician born and raised in Buenos Aires. He obtained a degree in mathematics in 2005 and a PhD in mathematics in 2009, both from the University of Buenos Aires, and worked as a professor for many years. He has published several peer-reviewed academic articles and has given more than 15 conference presentations.

In February 2022, he started consulting in the Web3 space, applying his mathematical knowledge and expertise to the cutting-edge field of cryptocurrencies and Decentralized Finance, and he has already become an expert in automated market makers.After finding out that the literature on automated market makers is frequently incomplete and not precise, he decided to write this book with Peter and Jesper so as to provide newcomers with a friendly, clear, and thorough explanation of the primary automated market makers and to offer developers a solid, broad, and profound exposition that will allow them to gain a deeper understanding in order to continue building.

**Peter Johannes Steffensen** is a mathematician holding a PhD in mathematics from Aarhus University. Since then, he has been working as a software architect in different financial institutions and has also had part-time university jobs. He has worked as a consultant in the Web3 space for over a year and saw there the perfect possibility to apply his mathematical knowledge and experience from the financial industry. His motivation for writing this book is to explain how the leading automated market makers work and to provide a single point of entry for people that look for information on this topic.

## ABOUT THE AUTHORS

**Jesper Kristensen** has worked in the Web3 space for years in both consulting and full-time capacities. He holds a PhD and two master's degrees in applied and engineering physics from Cornell University. Jesper worked four years at General Electric Research, publishing nine articles, one book chapter, and giving seventeen conference presentations and invited talks. Following this, he did stints in the tech industry and finance for three years and has led a large Web3 Research department. His primary motivation for writing this book is twofold. First, when Jesper was first introduced to the Web3 space, he noticed that a lot of confusion about how automated market makers worked circulated and that multiple people gave incorrect answers around their inner workings—and this book seeks to settle these types of conflicts. Second, he wants to offer a one-stop complete guide to learn about automated market makers.

# About This Book

Automated market makers (AMMs) are the underlying protocols used by decentralized exchanges (DEXs) to allow users to perform trades of cryptocurrencies in a decentralized way with no middlemen. In order books, a buy order must be matched with a sell order for a trade to occur. In other words, a coincidence of wants needs to happen, which is not always guaranteed, and someone needs to step in to make the market. But in the case of an AMM, users trade against the protocol itself, a smart contract—or rather a community pool of funds—which is, in simple terms, an open source computer program in which mathematical formulae define prices. The crucial point is that no banks or intermediaries charging exuberant fees are necessary, giving financial access to a much larger crowd, including less privileged people.

In this book, we perform a thorough study and a clear-cut exposition of the principal AMM designs (Uniswap v2, Uniswap v3, Balancer, and Curve), giving a detailed and in-depth description of how they work, unveiling the mathematics behind them and showing plenty of examples as well as novel proofs for several interesting facts. While highlighting the ideas and mathematics behind the several formulae, we aim for an instructive, precise, and thorough exposition. In this way, we intend that readers can fully understand how AMMs and DEXs work, find out their subtleties, grasp the formulae applied when trading and depositing or removing liquidity, and understand their logic and the mathematical aspects behind them. In each chapter, we describe each AMM's core idea. We derive the mathematical formulae used in the code, explaining the logical thinking involved and detailing the steps needed to reach those formulae. All these explanations are complemented with thoughtfully chosen examples that further help enlighten the readers on how the formulae work. In addition, the relevant parts of the codes of the AMMs are shown after explaining each formula so that the actual implementation of the formulae can be seen. Technicalities in this space take a while to figure out independently and often require knowing the Solidity programming language. Our book fulfills this need since we will detail how the mathematics of AMMs makes sense and how it couples to the code. We are convinced that sharing is vital and at the core of the vision of true decentralization.

This book is ideal for newcomers who want to learn about AMMs from scratch and is also suitable for experienced developers who want to delve deeper into the core features of these platforms or use this book as a detailed reference guide. And since we provide mathematical arguments and show the actual implementation of the formulae in the corresponding smart contracts, developers will benefit from both a theoretical and practical understanding of AMMs.

The book is organized as follows.

In the first chapter, we explain several preliminary concepts that readers need to know before diving into the following chapters so that we pave the way for newcomers to go through the book. We also give a historical overview of AMMs and conduct a broad literature review.

In the second chapter, we study the Uniswap v2 AMM in depth. This study will not only serve to understand how Uniswap v2 works but will also lay the foundations for our journey through the other AMMs. Explicitly, we explain how trades are performed in the Uniswap v2 AMM, how prices are computed, and how fees are charged, and we analyze how the price changes when a trade occurs. Several graphs will help the reader to visualize all these processes. We then show how liquidity can be deposited and how liquidity providers can remove the liquidity they have deposited. We also describe how liquidity providers earn trading fees, and we explain the risk they have of facing impermanent losses.

In the third chapter, we give an in-depth analysis of the Balancer AMM design. First, we show how the spot price and trading formulae are obtained. Then we explain how liquidity can be deposited and withdrawn. We analyze the single-asset and all-asset cases for both deposits and withdrawals, presenting some novel results with their respective proofs. At the end of the third chapter, we explain how Balancer pool tokens can be priced and give some examples.

In the fourth chapter, we perform a detailed study of the first version of the Curve AMM, called StableSwap, designed to make trades between pegged assets, such as stablecoins. We start Chapter 4 by explaining how the StableSwap formula is obtained, and we then give mathematical proofs that show that this design indeed works correctly. It is worth mentioning that these proofs cannot be found elsewhere, not even in the corresponding whitepaper. After that, we describe how trades are performed and how the necessary computations for trading assets are carried out. Then we explain how all-asset deposits, all-asset withdrawals, and single-asset withdrawals work. We end Chapter 4 by showing how to price StableSwap pool tokens.

In the last chapter, we thoroughly explain how the Uniswap v3 AMM works. Uniswap v3 implements concentrated liquidity, which, from the liquidity provider's perspective, makes it very different from the other AMMs described in this book's previous chapters. Thus, we start Chapter 5 by introducing the new concepts that appear with this new design. We then analyze how a liquidity provider's position varies when trades are performed and the price changes. After that, we study impermanent losses in Uniswap v3 and compare them with those of Uniswap v2. Then, we describe in detail how the Uniswap v3 AMM is implemented, showing the relevant parts of the smart contract code and the corresponding explanations. We end the book's last chapter with a detailed analysis of liquidity provisioning in Uniswap v3. We explain the concept of capital efficiency and prove that the performance of a Uniswap v3 position only depends on the parameters of the position and not on the other positions that might exist in the same liquidity pool.

We hope you find this book instructive and helpful and enjoy the journey.

# CHAPTER 1

# Introduction to AMMs

In this chapter, we will introduce many fundamental concepts newcomers need to know to pleasantly stroll through the pages of this book. We will also give a historical overview of automated market makers and round it off with a broad literature review that might help the interested reader delve deeper into related topics.

## 1.1 Preliminary Notions

In this section, we will take a quick tour around several basic concepts regarding blockchains and trading that are needed to better understand the contents of this book. Readers already familiar with the Ethereum mainnet and with standard trading terms may skip this section.

We think it makes sense to start this tour from the very beginning, that is, with the definition of a blockchain. A **blockchain** is a decentralized, distributed, and public digital ledger (or, more generally, database) that exists across a network and consists of a large number of pieces called **blocks** that store the information and have a specific storage capacity. When a block is completed, which happens every 15 seconds on average, it is closed and linked to the previously closed block, forming a chain of blocks from which the name of the blockchain is derived.

In a very simplistic way, we can think of a blockchain as a book in which the pages are the blocks. When a new page is completed, it is added to the end of the book with the corresponding reference to the previous page. In books, this reference can be the page numbering, but in

© Miguel Ottina, Peter Johannes Steffensen, Jesper Kristensen 2023
M. Ottina et al., *Automated Market Makers*, https://doi.org/10.1007/978-1-4842-8616-6_1

blockchains, it is not that simple. Blocks are connected cryptographically via so-called hash pointers. Since the hash depends on the entire content of the current block and includes the thumbnail from the previous block, the blockchain is considered tamper-proof. To keep this discussion simple, we will ignore sophisticated details about the blockchain technology such as temporary forks, limitations on the block size, and confirmations and refer the interested reader to other excellent resources on those topics, such as [7], [40], and [37]. As alluded to earlier, how the blocks are linked makes it impossible for anyone to modify their contents. Hence, blockchains provide a safe way to store financial information.

A proof-of-work blockchain is maintained secure and operational by **miners**, who are users that verify the legitimacy of the transactions that are going to be written in a new block and stored on the blockchain. For doing so, they earn fees, known as transaction fees or **gas fees**. More explicitly, users pay gas fees to miners for recording their transactions on the blockchain. Note that for a simple transaction of money, for example, between two parties, the amount of gas fee that has to be paid does not depend on the amount of the transaction that has to be recorded— transferring 100 units costs the same as transferring 1,000,000 units—but rather on the network congestion at the moment of the transaction. If a large number of users want their transaction to be recorded around the same time, they will offer to pay more gas fees so that their transaction gets prioritized by the miners, and hence, they do not have to wait. Sometimes, gas fees get too high, which constitutes a problem for users that want to make small transactions. This problem has been addressed in several ways, one of them being the creation of layer 2 networks on top of the main blockchain. **Layer 2 networks**—or simply L2 networks— are separate blockchains that extend the principal one and inherit the security guarantees that the principal blockchain offers. In a few words, a layer 2 network processes many transactions fast and cheaply. It regularly communicates with the main blockchain submitting bundles of transactions so that they get recorded on the principal blockchain, which is called **layer 1** (L1). For example, Polygon, Optimism, and Arbitrum are L2 networks built on the Ethereum blockchain.

In this book, we will focus on the Ethereum blockchain, also known as Ethereum network or Ethereum mainnet. In contrast to the Bitcoin blockchain, the Ethereum blockchain supports the creation, deployment, and functioning of smart contracts. A **smart contract** is a computer program that runs on a blockchain. Smart contracts can be thought of as contracts in which the rules or terms of the agreement are written in the computer code of the contract. Smart contracts' most common programming languages include Solidity, Rust, and Vyper. Smart contracts have many different uses, but in this book, we will focus on the smart contracts that allow the exchange of an amount of one asset for an amount of another one.

The assets that we will be mainly interested in are ERC-20 tokens. In the Ethereum network, **tokens** are blockchain-based digital assets that can be bought, sold, or traded, while **ERC-20** is a standard used within the Ethereum blockchain for fungible tokens. ERC stands for Ethereum Request for Comments, and ERC-20 was developed to ensure a consistent interface of tokens in Ethereum to standardize the ways they are defined, created, and transferred between accounts. We point out, though, that the blockchain's native currency—the one in which miners are paid—is not a token but a coin. For example, Ether, the native currency of the Ethereum network, is a coin and not an ERC-20 token. However, for simplicity, in the following chapters, we will use the word token to refer to either a token or a coin.

It is worth mentioning that a type $T$ of tokens is **fungible** if any certain amount $a$ of token $T$ has the same value as another amount $a$ of the same token $T$, and hence, those amounts are treated in the same way and can be exchanged for each other. For example, USDC (USD Coin) is an ERC-20 token—hence, a fungible token—and thus, 1 USDC is completely interchangeable for any other 1 USDC, and any amount of 1 USDC has the same value.[1] Other ERC-20 tokens are USDT, DAI, MATIC, LINK, and WBTC.

On the other hand, the famous NFTs are **nonfungible tokens**, that is, digitally unique tokens of the Ethereum blockchain. NFTs are a way to represent something unique in the form of an Ethereum-based asset. Since NFTs are tokens, they can be created, bought, sold, and traded. Most NFTs are built using a consistent standard known as ERC-721.

The places where blockchain assets are traded are called **exchanges**, and they can be either centralized or decentralized. **Centralized exchanges**—CEXs, for short—are companies that facilitate trading between users acting as a trusted third party. Examples of CEXs are Binance, Coinbase, and Kraken. CEXs usually implement **order books**, which are simply a detailed list of the prices at which the users are willing to buy or sell a particular asset, together with the amount of that asset they want to buy or sell. It is essential to point out that the trades that occur within a CEX are not recorded on the blockchain. Hence, users do not have to pay gas fees, making trades much cheaper and faster. Usually, CEXs charge users with a minimal **trading fee**—generally 0.1% of the amount of the trade, or even less than that. Another advantage of CEXs is that they usually enjoy much higher **liquidity** than their decentralized counterparts, meaning they have a larger amount of funds to trade. This is a significant advantage for traders since they can get better asset prices. Higher liquidity also implies a lower **bid/ask spread**, a smaller difference between the price at which an asset can be bought and the price at which it can be sold. In addition, CEXs usually have huge **trading volumes**—this is the amount of a particular pair of assets that is traded in a certain amount of time—with hundreds of transactions taking place each minute for each of the most common pairs of assets. Large trading volumes imply that traders have to wait very little time for their orders to go through if they are correctly placed.

---

[1] Paper money is another example of a fungible asset: a 100 USD banknote is equivalent to any other 100 USD banknote.

A huge disadvantage of CEXs is that the users do not fully own the assets they deposit into them. Although users rarely have any problem when dealing with the bigger and more renowned CEXs, it happened a few times that smaller CEXs disappeared, and their users could never recover their funds. This became painfully clear in the recent FTX[2] catastrophy.

On the other hand, **decentralized exchanges**—DEXs, for short—are blockchain-based apps that allow trading by means of smart contracts. Trades performed within a DEX are recorded on the blockchain and are simply transfers between the blockchain addresses of the two users involved in the trade. Thus, users keep full ownership of their assets at every moment.

Since DEXs work by means of smart contracts, the implementation of an order book in a DEX has an elevated gas cost because every modification of the order book—creation of a new order, cancellation of an existing order, or partial or total fulfillment of an order—will require a gas fee to be paid. Therefore, on-chain order books are not used within DEXs on L1 networks. We have to mention, though, that there exist some examples of off-chain order books with on-chain registration of trades, such as 1inch[3], Sushiswap[4], and dYdX[5]. However, in L1 networks, trades are usually made by means of **automated market makers**—AMMs, for short—which are smart contracts that perform trades using a certain amount of funds that is available to each of them. In this way, a trader does not need to find a counterpart to perform a trade but rather interacts with the smart contract to buy or sell assets. The place in which the funds of the AMM are stored is called liquidity pool. Funds are deposited into **liquidity pools**

---

[2] https://www.forbes.com/sites/jonathanponciano/2022/11/11/ftx-files-for-bankruptcy-former-billionaire-sam-bankman-fried-resigns-as-ceo/?sh=7353734231d9

[3] https://app.1inch.io/

[4] https://app.sushi.com/limit-order

[5] https://trade.dydx.exchange/trade/

by other users called **liquidity providers**, who will earn trading fees from each trade performed within the AMM they have provided liquidity to. The amount of assets that a liquidity pool holds is called **liquidity** or **depth** of the pool. As we shall see, a pool with more liquidity can offer better conditions for traders.

# 1.2  Historical Overview

The appearance of AMMs followed the need to implement on-chain trading efficiently. Initially, DEXs used order books to allow trading of assets. However, on-chain order books have several significant drawbacks. First of all, as we mentioned in the previous section, each order must be registered on-chain, which has an elevated gas cost (as we mentioned before, in order to add or remove a limit order, a gas fee must be paid). Secondly, low liquidity implies a significant bid-ask spread and a slow execution of orders for pairs of assets that do not have a high trading volume.

To address these problems, Vitalik Buterin proposed in 2016 the implementation of on-chain AMMs [9], taking ideas from prediction markets [18]. Buterin's idea was first applied to AMMs in 2017 in the Bancor protocol [21], which allowed trades between the Bancor Network Token (BNT)—the protocol token—and other ERC-20 tokens through specific smart tokens. Afterward, Martin Koppelmann suggested a simplification to Buterin's ideas and came up with the constant product formula [9]. And several months later, in November 2018, Uniswap v1 was launched with the implementation of the constant product formula and the liquidity pools, and allowing trades between ETH and any other ERC-20 token.

In January 2020, Curve Finance[6] launched a new AMM that was designed exclusively to trade between stablecoins, with low price impact and high efficiency [11]. In March 2020, the Balancer[7] protocol [27] was deployed on the ETH mainnet, introducing a geometric mean formula that allowed the creation of liquidity pools consisting of up to 8 ERC-20 tokens. Shortly after, in May 2020, Uniswap v2[8] was launched, incorporating liquidity pools consisting of any pair of ERC-20 tokens and thus removing the necessity of the ETH bridge of Uniswap v1 [1]. Following these ideas, many other DEXs were deployed during 2020 on other networks, for example, Sushiswap[9] (on Polygon), PancakeSwap[10] (on Binance Smart Chain), and DODO[11] (on the Ethereum mainnet, Polygon, and Binance Smart Chain, among others). In particular, DODO introduced an innovative Proactive Market Maker algorithm to provide efficient on-chain liquidity.

In October 2020, Bancor v2.1[12] was introduced, offering impermanent loss protection to long-term liquidity providers. Due to the impermanent loss protection, each token pool had an upper limit determined by Bancor DAO's[13] co-investment.

Several months later, in May 2021, Uniswap v3[14] appeared and started offering an entirely new concept to liquidity providers: concentrated liquidity. This means that liquidity providers can deposit liquidity in a chosen price range, which makes Uniswap v3 very different from Uniswap v2. Later, both versions of Uniswap were also deployed on Polygon, Optimism, and Arbitrum.

---

[6] https://curve.fi/

[7] https://balancer.fi/

[8] https://uniswap.org/

[9] https://sushi.com/

[10] https://pancakeswap.finance/

[11] https://app.dodoex.io/

[12] www.bancor.network/

[13] DAO stands for Decentralized Autonomous Organization.

[14] https://docs.uniswap.org/protocol/introduction

In June 2021, the Clipper[15] DEX was launched. Clipper is designed to have the lowest per-transaction costs for small to medium-sized trades. In this way, it intends to be the best place for self-made traders to buy and sell the most popular cryptoassets.

In November 2021, the Lifinity[16] protocol was introduced. Lifinity is deployed on Solana's network and applies leverage to liquidity and a proactive market maker design, using Pyth Network as an oracle, in order to improve capital efficiency and reduce impermanent loss.

Some months later, in March 2022, Platypus[17] was deployed on the Avalanche blockchain. Platypus is designed for exchanging stablecoins with much lower price impact than other similar AMMs.

In May 2022, Bancor 3[18] was launched. In Bancor 3, there is a single pool to stake BNT and earn yield from the entire network. In addition, all trades can be performed in a single transaction—they no longer require transfers via the protocol token—and there are no deposit limits on Bancor liquidity pools. Bancor 3 offers full impermanent loss protection from day one of the deposit.

In June 2022, the beta version of the Swaap Finance[19] AMM was launched on Polygon. The Swaap protocol is based on Balancer's AMM and is designed to almost get rid of liquidity providers' impermanent losses. This is achieved by tracking prices from oracles and implementing a dynamic and asymmetric spread.

DEXs experienced a significant growth in 2021, having recorded more than 1 trillion USD in annual trading volume—which is more than an 800%

---

[15] https://clipper.exchange/
[16] https://lifinity.io/
[17] https://platypus.finance/
[18] https://home.bancor.network/
[19] www.swaap.finance/

increase from 2020—and with a total value locked above 200 billion USD at the start of 2022,[20] growing from 20 billion USD a year before. Moreover, new Decentralized Finance (DeFi) products are continuously developed aiming to enable people to do most of the things that financial institutions allow them to do, such as trading, borrowing and lending assets, earning interest through deposits, and buying or selling insurance, but doing it faster and without paperwork or intermediaries in a very secure way.

In this line, the research on AMMs has also been flourishing since the beginning of 2021. Many articles have appeared covering diverse topics, ranging from modifications, extensions, and generalizations of well-known AMMs to analyses of current problems of AMMs and proposals of possible solutions, and from expository articles to more analytical ones.

Regarding expository articles, in [29], Mohan presents a unified framework to characterize different types of AMMs that are currently popular as DEXs. He also provides an intuitive geometric representation of how an AMM operates and delineates the similarities and differences across the various types of AMMs. In [39], Xu, Paruch, Cousaert, and Feng systematize the knowledge around AMM-based DEXs and compare the top AMM protocols' mechanics, illustrating their conservation functions, as well as price impact and divergence loss functions. In addition, [32] provides a brief overview of DEXs and automated market makers and discusses the trade-off between security and market efficiency. Also, a comprehensive overview of several of the main AMMs, together with a discussion on impermanent loss, concentrated liquidity management, and just-in-time liquidity, can be found in [17].

Other works focus on the analysis of the existent AMMs. For example, in [6], Angeris, Kao, Chiang, Noyes, and Chitra give a formal study of constant product markets and their generalizations and numerically demonstrate via simulations that Uniswap is stable under a wide range of market conditions. On the other hand, [16] examines how the

---

[20] https://defillama.com/

introduction of concentrated liquidity has changed the liquidity provision market in automated market makers such as Uniswap by comparing average liquidity provider returns from trading fees before and after its introduction. Also, in [13], Elsts shows how to derive some of the results of the Uniswap v3 whitepaper ([2]), presents several other equations that are not discussed in the whitepaper, and shows how to apply these equations. Finally, [35] compares mathematical models for automated market makers, including logarithmic market scoring rule (LMSR), liquidity-sensitive LMSR (LS-LMSR), constant product, mean and sum AMMs, and others.

Remarkably, a great amount of work is devoted to generalizing, extending, and improving existing AMMs and developing new products based on them. For example, in [36], Wang describes in great detail the actual implementation of an AMM for liquidity pools consisting of two tokens, whose constant curve is an arc of an ellipse. This AMM is implemented on CoinSwap.[21] In [31], a new formula for automated liquidity provision—called the constant power sum invariant—is introduced. This formula incorporates time to maturity as input and ensures that the liquidity pool offers a constant interest rate for a given ratio of its reserves. The same formula but with the exponents being equal to a parameter rather than to the time to maturity is studied in detail in [38]. In addition, [14] proposes a mathematical model for AMM composition, defining sequential and parallel composition operators for AMMs, and [41] describes and studies an AMM design that is obtained as a nesting of other AMMs. As we can see, the last two articles show how different formulae for AMMs can be combined to obtain new AMMs with different properties.

Another line of research focuses on using AMMs to replicate payoff functions. For example, Evans shows in [15] how liquidity providers' shares can be used to replicate target payoffs in geometric mean market makers

---

[21] http://coinswapapp.io/

with time-varying and potentially stochastic weights. In addition, in [4], Angeris, Evans, and Chitra present a method for constructing Constant Function Market Makers (CFMMs) whose portfolio value functions match a desired payoff. Furthermore, in [5], they show that any monotonic payoff can be replicated using only liquidity provider shares in CFMMs, without the need for additional collateral or oracles.

Another important research topic is studying and reducing the weaknesses of AMMs, which include frontrunning, transaction reordering, and sandwich attacks. The three of them consist of a party that intends to rearrange the order in which the transactions will be performed in order to earn a profit. Several works propose possible ways to alleviate these problems [42, 19, 28]. In addition, in [20], Heimbach and Wattenhofer categorize and analyze state-of-the-art transaction reordering manipulation mitigation schemes and evaluate their full impact on the blockchain. By doing so, they provide a complete picture of the strengths and weaknesses of current mitigation schemes and discover that currently no scheme fully meets all the demands of the blockchain ecosystem.

It is also essential to understand, analyze, and propose solutions to lessen market inefficiencies in AMMs, such as the availability of cyclic arbitrage opportunities [8]. Additionally, the impermanent loss—also called divergence loss—can be thought of as a kind of market inefficiency. And since the impermanent loss is a loss that liquidity providers can face due to deviations in the price of the assets, the analysis of impermanent losses [3, 25, 34, 22] is meaningful for liquidity providers.

On the other hand, some works focus on making AMMs more dynamic. For example, [30] proposes a novel approach called Dynamic Automated Market Making to allow both flexible fee adjustment and dynamic pricing curve setups. This helps reduce the impact of impermanent loss due to its flexible fee approach while allowing better capital efficiency using different pricing curve setups tailored for the pair of tokens in each pool. Also, in [23], Krishnamachari, Feng, and Grippo propose a new approach to the design of AMMs with the idea of dynamic curves, which utilizes input from a market price oracle to modify the mathematical relationship between the assets so that the pool price is continuously and automatically adjusted to make it identical to the market price, thus eliminating arbitrage opportunities.

As we can see, AMMs are at the cutting edge of DeFi, not only from the practical point of view but also from the theoretical one. The rapid growth this field is experiencing poses a challenge to everyone in this ecosystem, not just to keep pace with the new developments but, more importantly, to ensure that everyone can participate and nobody is left behind. Since DeFi is an open and global financial system and the Web3 technology is meant to be decentralized, everybody must have options to understand how these technologies work so that they can take part in them if they want to do so.

# 1.3 Summary

In this chapter, we introduced several fundamental concepts such as fungible and nonfungible tokens, trading fees, smart contracts, liquidity pools, and liquidity providers. In addition, we gave a historical account of automated market makers and a broad description of related research lines.

In the next chapter, we will study the Uniswap v2 AMM in detail, which will serve as the starting point of our journey through the main AMMs' designs.

# CHAPTER 2

# Uniswap v2

Uniswap v2 is a decentralized exchange based on a system of smart contracts on the Ethereum blockchain and other networks (such as Polygon, Optimism, Arbitrum). It is formed by liquidity pools that enable automated market making; that is, they enable traders to buy and sell assets against the protocol without the need for a third party. Each Uniswap v2 liquidity pool consists of reserves of two ERC-20 tokens deposited by liquidity providers, who benefit from the fees that the protocol charges to traders. The collected fees are shared proportionally among all liquidity providers.

Uniswap v2 liquidity pools are based on a product formula that considers the amount of reserves of each of the two tokens of the pool. This product formula plays a crucial role in determining the amounts of each token involved in any trade. Consequently, the product formula determines the price of one of the pool tokens in terms of the other. Moreover, the price of an asset in a Uniswap v2 liquidity pool follows the actual market price of that asset due to the actions of external arbitrageurs, who detect price inconsistencies between the market price and the internal price of the liquidity pool and make a trade in order to gain an instant benefit from the difference between the said prices, making the internal price of the oracle once again equal to the market price.

In this chapter, we will explain in detail how the Uniswap v2 AMM works and delve into the mathematical concepts behind it. We will explain how the trading formulae and the spot price formula are derived from the product formula and show how liquidity providers can deposit and remove

© Miguel Ottina, Peter Johannes Steffensen, Jesper Kristensen 2023
M. Ottina et al., *Automated Market Makers*, https://doi.org/10.1007/978-1-4842-8616-6_2

liquidity from the liquidity pools. In the process, we will analyze several interesting questions and give various illustrative examples.

# 2.1 Trading in a Uniswap v2 pool

Consider a Uniswap v2 liquidity pool formed by two tokens, $X$ and $Y$. Let $x$ and $y$ be variables representing the amounts of tokens $X$ and $Y$, respectively, that are in the pool at a particular moment. As we mentioned before, Uniswap v2 liquidity pools are based on a *constant product formula*. This means that the pool balances $x$ and $y$ satisfy the following equation:

$$x \cdot y = L^2, \tag{2.1}$$

where $L$ is a positive number that is called *liquidity parameter* of the pool. For the moment, we will assume that $L$ is a constant value. We will see later how the liquidity parameter $L$ can change.

In order to see how the constant product formula works, consider a Uniswap v2 liquidity pool with no fees whose reserves are 100 ETH and 400,000 USDC. From the constant product formula, we obtain that $100 \cdot 400{,}000 = L^2$. Hence, $L^2 = 40{,}000{,}000$. Suppose that a trader wants to buy 20 ETH. To do so, they must deposit a certain amount of USDC into the pool. Observe that if the pool sends 20 ETH to the trader, there will be 80 ETH left in the pool. From the constant product formula, we obtain that the balance $B$ of USDC when there are 80 ETH left in the pool has to satisfy

$$80 \cdot B = L^2 = 40{,}000{,}000.$$

Thus, $B = 5{,}000{,}000$. That is, if there are 80 ETH in the pool, then there must be 500,000 USDC in the pool. Therefore, the trader will have to deposit an amount of $500{,}000 - 400{,}000 = 100{,}000$ USDC so as to be given 20 ETH. In other terms, the pool balances before and after the trade must satisfy Equation 2.1, and thus they can be regarded as points of the real plane that belong to the curve defined by Equation 2.1, as Figure 2-1 shows.

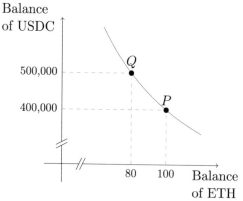

$P = (100, 400000)$: Pool state before the trade.
$Q = (80, 500000)$: Pool state after the trade.

**Figure 2-1.** *Pool states before and after the trade in a pool with no fees*

We will now analyze how the constant product formula works in the general case in order to obtain several important formulae that will be used throughout this chapter. We will assume first that the liquidity pool has no fees.

Suppose that a trader wants to buy an amount $a$ of token $X$. To do so, they must deposit an amount $b$ of token $Y$. Let $A$ and $B$ be the balances in the pool of tokens $X$ and $Y$, respectively, immediately before the trade. Then,

$$AB = L^2.$$

After the trade, the balance of token $X$ in the pool will be $A - a$, and the balance of token $Y$ in the pool will be $B + b$. Thus,

$$(A-a)(B+b) = L^2.$$

Hence,

$$L^2 = (A-a)(B+b) = AB + Ab - aB - ab = L^2 + Ab - aB - ab$$

and thus,

$$Ab - aB - ab = 0.$$

Therefore, we can isolate $b$ to obtain

$$b = \frac{aB}{A-a}. \qquad (2.2)$$

This means that in order to receive an amount $a$ of token $X$, the trader must deposit an amount $\dfrac{aB}{A-a}$ of token $Y$.

Similarly, we can isolate $a$ to obtain

$$a = \frac{Ab}{B+b}. \qquad (2.3)$$

This means that if the trader deposits an amount $b$ of token $Y$, they will receive an amount $\dfrac{Ab}{B+b}$ of token $X$.

## 2.1.1 Spot Price

A fundamental concept is that of the *spot price*. The spot price of token $X$ in terms of token $Y$ is defined as the price we pay per token $X$ that we receive after depositing an infinitely small amount of token $Y$. In order to study the spot price and obtain a formula to compute it, we need to give a more formal definition.

Let $A$ and $B$ be the balances in the pool of tokens $X$ and $Y$, respectively. Suppose that a trader deposits an amount $b$ of token $Y$ and receives an amount $a$ of token $X$. Then, the price that the trader paid for each unit of token $X$ is $\dfrac{b}{a}$. We also say that $\dfrac{b}{a}$ is the *effective price* (of token $X$ in terms of token $Y$) paid by the trader.

Let $p$ be the spot price of token $X$ in terms of token $Y$ and let $p_e(b)$ be the effective price (of token $X$ in terms of token $Y$) paid by the trader when depositing an amount $b$ of token $Y$.

Then, the *spot price* is defined as

$$p = \lim_{b \to 0} p_e(b) = \lim_{b \to 0} \frac{b}{a}$$

(recall that $a$ depends on $b$).

Applying Equation 2.3, we obtain that

$$p_e(b) = \frac{b}{a} = \frac{b}{\dfrac{Ab}{B+b}} = \frac{b(B+b)}{Ab} = \frac{B+b}{A} = \frac{B}{A} + \frac{b}{A}.$$

Thus,

$$p = \lim_{b \to 0} p_e(b) = \lim_{b \to 0} \left( \frac{B}{A} + \frac{b}{A} \right) = \frac{B}{A}. \tag{2.4}$$

Hence, the spot price when the pool state is $(A, B)$ is equal to $\dfrac{B}{A}$, which coincides with the slope of the line that passes through the origin $(0, 0)$ and the point $(A, B)$.

Note also that since $b > 0$,

$$p_e(b) = \frac{B}{A} + \frac{b}{A} > \frac{B}{A} = p,$$

that is, the effective price paid by the trader is greater than the spot price. Note also that the difference between them is

$$p_e(b) - p = \frac{b}{A}, \tag{2.5}$$

which is small if the deposited amount $b$ is small with respect to the balance $A$ of token $X$. This means that for pools with large liquidity, this difference will be small.

The spot price can then be interpreted as the price the liquidity pool offers traders at a particular moment. However, as we have seen, traders will always have to pay more than that. The difference between the amount $b$ that a trader pays (or deposits) and the amount that the trader would have paid if they had bought an amount $a$ of the asset at a price equal to the spot price $p$ is called *price impact*[1] and is given by

$$\text{Price impact} = b - ap.$$

Note that

$$\text{Price impact} = a\left(\frac{b}{a} - p\right) = a\left(p_e(b) - p\right) = \frac{ab}{A}$$

by Equation 2.5.

In Figure 2-2, we show the geometric interpretations of both the spot price and the price impact. With the previous notations, we mentioned before that when the pool state is $(A, B)$, the spot price $p$ is equal to $\frac{B}{A}$, which coincides with the slope of the line segment that goes from $(0,0)$ to

---

[1] We point out that price impact is called slippage by some authors. Properly speaking, slippage is the difference between the quoted price of a trade and the final execution price that is caused by external market movements. We refer the interested reader to https://integral.link/post/slippage-and-price-impact-in-defi-explained for more details.

$(A, B)$ (this is the dotted segment of Figure 2-2). Note also that if the state of the pool before a trade is $P = (A, B)$ and the state of the pool after that trade is $P' = (A', B') = (A - a, B + b)$, then the effective price that the trader paid for each unit of token $X$ is

$$p_e = \frac{B' - B}{A - A'}.$$

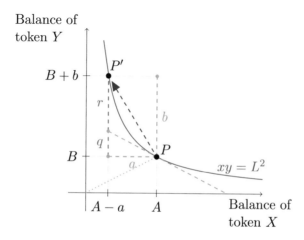

Segment from $(0,0)$ to $P$. Its slope is $\frac{B}{A}$, that is, the spot price at $P$.

Portion of the tangent line to the curve defined by $xy = L^2$ at $P$. Its slope is $-\frac{B}{A}$, which is the opposite of the spot price at $P$.

$b$: Amount of token $Y$ deposited.

$a$: Amount of token $X$ received.

$q$: Amount of token $Y$ that would have been paid if the effective price had been equal to the spot price.

$r$: Price impact (equal to $b - q$).

***Figure 2-2.*** *Geometric interpretation of the spot price and the price impact*

Since the point $(A', B')$ belongs to the curve $xy = L^2$, we can regard $B'$ as a function of $A'$, and hence, $p_e$ can also be thought of as a function of $A'$ (recall that $A$ and $B$ are fixed). Hence, we obtain that the spot price $p$ is given by

$$p = \lim_{A' \to A} p_e(A') = \lim_{A' \to A} \frac{B' - B}{A - A'} = -\lim_{A' \to A} \frac{B' - B}{A' - A},$$

which is the opposite of the slope of the tangent line to the curve $xy = L^2$ at the point $(A, B)$. Note that this slope can also be computed by isolating $y$ from the equation $xy = L^2$ to obtain

$$y = \frac{L^2}{x},$$

from where we get that the derivative of $y$ with respect to $x$ is

$$y' = -\frac{L^2}{x^2},$$

and thus, the slope of the tangent line to the curve $xy = L^2$ at the point $(A, B)$ is

$$-\frac{L^2}{A^2} = -\frac{AB}{A^2} = -\frac{B}{A} = -p,$$

as expected. We can also see this in Figure 2-2.

Regarding the price impact, since the slope of the tangent line is the opposite of the spot price $p$, it follows that in Figure 2-2, $q = ap$, and hence, $q$ is the amount of token $Y$ that would have been paid if the whole trade had been executed at the spot price. Therefore, we obtain that the price impact $s$ is equal to $b - q$, and hence, it can be interpreted as the length of the dashed segment labelled $r$ of Figure 2-2.

**Example 2.1.** Consider a Uniswap v2 liquidity pool with no fees that has 200 ETH and 800,000 USDC. Note that the spot price of ETH in terms of USDC is

$$p = \frac{800,000}{200} = 4,000.$$

Suppose that a trader deposits 200,000 USDC into the pool so as to receive ETH. Applying Equation 2.3, we obtain that the amount of ETH that the trader receives is

$$a = \frac{200 \cdot 200,000}{800,000 + 200,000} = 40.$$

Hence, the effective price paid by the trader is

$$\frac{200,000}{40} = 5,000,$$

which is more than the spot price $p$. This means that the trader paid 5,000 USDC for each ETH.

Since the trader bought 40 ETH and the spot price was 4,000, the trader would have expected to need to deposit 40×4,000 USDC, that is, 160,000 USDC, but instead, 200,000 USDC were required for the transaction. The difference

$$200,000 - 160,000 = 40,000 \text{ USDC}$$

is the price impact of the trade, which can be seen as a loss for the trader.

## 2.1.2 Accounting for Fees

Equations 2.2 and 2.3 are valid if the liquidity pool has no fees, but in general, traders are charged a fee for trading. The collected fees are added to the pool reserves and then paid to liquidity providers in a way that is proportional to the deposits they have made.

Uniswap v2 charges fees on the way in. That means that when a trader wants to buy or sell assets, first, the protocol charges the fee on the amount that the trader deposits, and then the remaining amount is actually traded using the previous arguments, as Figure 2-3 shows.

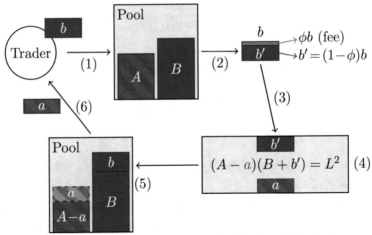

(1) The trader deposits an amount $b$ of token $Y$ into the pool.
(2) The AMM charges the fee on the amount $b$.
(3) The remaining amount $b' = (1 - \phi)b$ is actually traded.
(4) The AMM computes the amount $a$ of token $X$ that has to be given to the trader, following the curve described by $xy = L^2$ and using $b'$ as the incoming amount of token $Y$.
(5) The balance of token $X$ in the pool decreases by $a$, and the balance of token $Y$ in the pool increases by $b$.
(6) An amount $a$ of token $X$ is given to the trader.

**Figure 2-3.** *Trading mechanism with fees*

We will see how the mathematics for this works. Let $A$ and $B$ be the balances of tokens $X$ and $Y$ in a Uniswap v2 liquidity pool. Let $\phi \in [0, 1)$ be the pool fee. For example, if the pool fee is 0.3%, then $\phi = 0.003$. Let $L$ be the liquidity parameter of the pool. Thus, $AB = L^2$. Suppose that a trader deposits an amount $b$ of token $Y$ and receives an amount $a$ of token $X$. Since the fee is charged on the way in, we may assume that the deposited amount is $(1 - \phi)b$ and proceed as if the pool had no fees. Hence, the amounts $a$ and $b$ satisfy the following equation:

$$(A-a)(B+(1-\phi)b) = L^2.$$

Thus,

$$
\begin{aligned}
L^2 &= (A-a)(B+(1-\phi)b) \\
&= AB + A(1-\phi)b - aB - a(1-\phi)b \\
&= L^2 + A(1-\phi)b - aB - a(1-\phi)b
\end{aligned}
$$

and hence,

$$A(1-\phi)b - aB - a(1-\phi)b = 0. \tag{2.6}$$

Isolating $a$, we obtain that

$$a = \frac{A(1-\phi)b}{B+(1-\phi)b} \tag{2.7}$$

This means that if the trader deposits an amount $b$ of token $Y$, they will receive an amount $\dfrac{A(1-\phi)b}{B+(1-\phi)b}$ of token $X$.

Similarly, isolating $b$ from Equation 2.6, we obtain that

$$b = \frac{aB}{(1-\phi)(A-a)} \tag{2.8}$$

This means that in order to receive an amount $a$ of token $X$, the trader has to deposit an amount $\dfrac{aB}{(1-\phi)(A-a)}$ of token $Y$.

Equations 2.7 and 2.8 are covered by the portion of the Uniswap v2 code[2] given in Listing 2-1, applying the following translation between the variables of the code and our notation:

$$\begin{aligned}
\text{reserveIn} &= B, \\
\text{reserveOut} &= A, \\
\text{amountIn} &= b, \\
\text{amountOut} &= a,
\end{aligned}$$

Observe that the actual code (given in Listing 2-1) employs a trading fee of 0.3% and performs a multiplication by 1,000 that is cancelled when the quotient is computed.

***Listing 2-1.*** Functions getAmountOut and GetAmountIn of the smart contract of Uniswap v2

```
// given an input amount of an asset and pair
↪   reserves, returns the maximum output amount of the
↪   other asset
function getAmountOut(uint amountIn, uint reserveIn,
↪   uint reserveOut) internal pure returns (uint
↪   amountOut) {
    require(amountIn > 0, 'UniswapV2Library:
↪   INSUFFICIENT_INPUT_AMOUNT');
    require(reserveIn > 0 && reserveOut > 0,
↪   'UniswapV2Library: INSUFFICIENT_LIQUIDITY');
    uint amountInWithFee = amountIn.mul(997);
```

---

[2] https://github.com/Uniswap/v2-periphery/blob/master/contracts/libraries/UniswapV2Library.sol

```
    uint numerator = amountInWithFee.mul(reserveOut);
    uint denominator =
↪   reserveIn.mul(1000).add(amountInWithFee);
    amountOut = numerator / denominator;
}

// given an output amount of an asset and pair
↪   reserves, returns a required input amount of the
↪   other asset
function getAmountIn(uint amountOut, uint reserveIn,
↪   uint reserveOut) internal pure returns (uint
↪   amountIn) {
    require(amountOut > 0, 'UniswapV2Library:
↪   INSUFFICIENT_OUTPUT_AMOUNT');
require(reserveIn > 0 && reserveOut > 0,
↪   UniswapV2Library: INSUFFICIENT_LIQUIDITY');
    uint numerator =
↪   reserveIn.mul(amountOut).mul(1000);
    uint denominator =
↪   reserveOut.sub(amountOut).mul(997);
    amountIn = (numerator / denominator).add(1);
}
```

It is important to point out that after each transaction, the pool reserves are updated in the following way:

$$A' = A - a$$
$$B' = B + b.$$

Thus, the liquidity parameter is also updated to a value $L'$ such that $L' > 0$ and

$$(L')^2 = (A-a)(B+b) = AB + Ab - aB - ab = L^2 + (A-a)b - aB$$

and since from Equation 2.6 we have that $aB = (A - a)(1 - \phi)b$, we obtain that

$$\left(L'\right)^2 = L^2 + \phi b\left(A - a\right). \tag{2.9}$$

Note that $(L')^2 > L^2$ if $\phi > 0$. That is, if the liquidity pool has nonzero fees, then the liquidity parameter will increase a bit after each transaction.

Observe also that the spot price changes after the transaction. In this case, the new spot price is

$$p' = \frac{B + b}{A - a}.$$

In summary, if the liquidity pool has no fees, its liquidity parameter remains the same after each trade (this is $L' = L$ with the previous notations). In addition, if a trader deposits an amount $b$ of token $Y$ and receives an amount $a$ of token $X$, and as above, $A$ and $B$ are the pool balances before the trade and $A'$ and $B'$ are the pool balances after the trade, then the point $(A', B')$ is located on the same multiplicative inverse curve as the point $(A, B)$, as Figure 2-4.

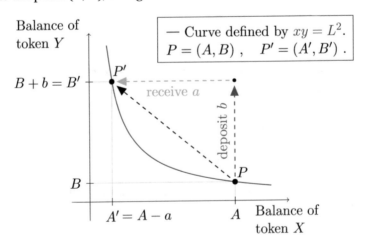

***Figure 2-4.*** *Change of the pool state when trading in a pool with no fees*

On the other hand, if the liquidity pool has a nonzero fee $\phi$, then the liquidity parameter increases after each trade (this is $L' > L$ with the previous notations). If the pool reserves before a trade are $A$ and $B$, and a trader deposits an amount $b$ of token $Y$ and receives an amount $a$ of token $X$, then

$$(A-a)(B+b(1-\phi))=L^2,$$

as we showed before. That is, the point $Q = (A - a, B + b(1 - \phi))$ is located on the curve defined by the liquidity parameter $L$, as we can see in Figure 2-5 (curve in solid line). Observe that the updated pool balances are defined by the point $P' = (A - a, B + b)$ and that $(A - a, B + b) = (A - a, B + b(1 - \phi)) + (0, b\phi)$. Thus, the point $P'$ is now located on a new multiplicative inverse curve (the dashed one in Figure 2-5).

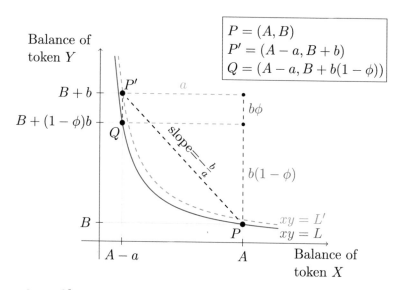

**Figure 2-5.** *Change of the pool state when trading in a pool with fees*

Note that the absolute value of the slope of the segment between the points $P = (A,B)$ and $P' = (A - a, B + b)$ shown in Figure 2-5 is exactly the deposited amount of token $Y$ per unit of token $X$, that is, the effective price paid by the trader.

# 2.2 Impact of the Trades on the Price

When a trade is executed, the balances of the tokens in the pool change, and as a consequence of the trade, the spot price also changes. In this section, we will perform several analyses to show how the spot price varies when trades occur.

## 2.2.1 A Simple Example

We will give first a simple example that shows how the spot price of a token increases after a trader purchases an amount of it.

**Example 2.2** (Shift in the average buy price). Consider a Uniswap v2 pool with ETH and USDC, having a fee of 0.3%. Suppose that the pool has reserves of 10,000 ETH and 40,000,000 USDC. Let $A = 10,000$ and $B = 40,000,000$. Note that the spot price is 4,000 USDC/ETH.

Suppose that a trader wants to buy an amount of 500 ETH from the pool. To find out the amount of USDC that the trader has to deposit in order to obtain 500 ETH, we apply Equation 2.8 with $a = 500$:

$$b_1 = \frac{a \cdot B}{(1-\phi) \cdot (A-a)} \approx 2{,}111{,}598 \ \text{USDC}.$$

Hence, the trader has to deposit approximately 2,111,598 USDC in order to obtain 500 ETH. This is equivalent to an average price of approximately 4,223 USDC/ETH. If the trader wants to buy 500 ETH again, we need to compute the updated balances (after the first trade):

$$A' = 9,500 \qquad (\text{balance of ETH}),$$
$$B' = 42,111,598 \quad (\text{balance of USDC})$$

Again, to compute the amount of USDC that the trader has to deposit in order to obtain 500 ETH, we apply Equation 2.8 with $a = 500$ and the new balances:

$$b_2 = \frac{a \cdot B'}{(1-\phi)\cdot(A'-a)} \approx 2,346,573 \quad \text{USDC}.$$

This means that the trader has to deposit approximately 2,346,573 USDC in order to obtain 500 ETH. This is equivalent to an average price of roughly 4,693 USDC/ETH. We find that the price of ETH in terms of USDC has increased due to the fact that the amounts of USDC and ETH have respectively increased and decreased reflecting a higher demand for ETH and thus a higher price for ETH in terms of USDC.

Observe that in general, and with the previous notations, if we buy an amount $a$ of token $X$, the total price that we have to pay is

$$b_1 = \frac{a \cdot B}{(1-\phi)\cdot(A-a)}.$$

If after that we buy an amount $a$ of token $X$ again, the new total price that we have to pay is

$$b_2 = \frac{a \cdot (B+b_1)}{(1-\phi)\cdot(A-2a)}.$$

Note that

$$b_1 = \frac{a \cdot B}{(1-\phi)\cdot(A-a)} < \frac{a \cdot (B+b_1)}{(1-\phi)\cdot(A-2a)} = b_2$$

since $B < B + b_1$ and $A - a > A - 2a$.

## 2.2.2 Analysis of Two Consecutive Trades

We will now extend the previous example to study if there is any difference between buying 500 ETH in only one trade and buying 500 ETH in two consecutive trades of 250 ETH each.

**Example 2.3** (ETH purchase price comparison). Consider again a liquidity pool of 10,000 ETH and 40,000,000 USDC. Let $A = 10,000$ and $B = 40,000,000$.

Suppose first that the pool has no fees. Using Equation 2.8, we obtain that in order to receive 500 ETH, we have to deposit

$$b_0 = \frac{a \cdot B}{A - a} = \frac{500 \cdot 40,000,000}{10,000 - 500} \approx 2,105,263 \ \text{USDC}.$$

On the other hand, for two consecutive purchases of 250 ETH each, we need to pay

$$b_1 = \frac{a \cdot B}{A - a} = \frac{250 \cdot 40,000,000}{10,000 - 250} \approx 1,025,641 \ \text{USDC and}$$

$$b_2 = \frac{a \cdot B'}{A' - a} = \frac{250 \cdot 41,025,641}{9,750 - 250} \approx 1,079,622 \ \text{USDC}$$

(note that the values of $A$ and $B$ are updated to $A' = 9,750$ and $B' = 41,025,641$ after the first purchase). We observe that

$$b_0 = b_1 + b_2,$$

that is, the amount of USDC that has to be paid in order to buy 500 ETH in one step is the same as the amount of USDC needed to buy 500 ETH in two consecutive steps of 250 ETH each.

This fact has an interesting graphical interpretation. Since the pool has no fees, the liquidity parameter $L$ remains constant, and thus, the points representing the different states of the pool belong to the same multiplicative inverse curve. Therefore, the point obtained after subtracting 500 from the ETH coordinate of the initial state is the same as the point obtained by subtracting 250 from the ETH coordinate of the initial state two times, as Figure 2-6 shows.

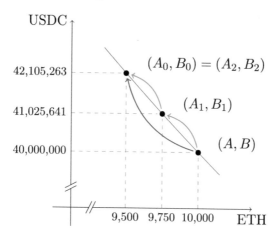

$(A, B)$: Initial pool state.
$(A_0, B_0)$: Pool state after one trade of 500 ETH.
$(A_1, B_1)$: Pool state after one trade of 250 ETH.
$(A_2, B_2)$: Pool state after two consecutive trades of 250 ETH each.

**Figure 2-6.** *Comparison between making a trade in one step and dividing it into two trades in a pool with no fees*

Now suppose that the pool has a fee of 0.3%. From the previous example, we see that in order to obtain 500 ETH, we have to pay approximately 2,111,598 USDC. On the other hand, for two consecutive buys of 250 ETH, we need to deposit the following amounts of USDC:

$$b_1 = \frac{a \cdot B}{(1-\phi) \cdot (A-a)} = \frac{250 \cdot 40,000,000}{0.997 \cdot (10,000 - 250)} \approx 1,028,727$$

$$b_2 = \frac{a \cdot (B+b_1)}{(1-\phi) \cdot (A-2a)} \approx \frac{250 \cdot 41,028,727}{0.997 \cdot (10,000 - 500)} \approx 1,082,952.$$

Hence, in order to buy 500 ETH in two steps, we need to pay approximately $1{,}028{,}727 + 1{,}082{,}952 = 2{,}111{,}679$ USDC, which is more than the 2,111,598 USDC needed to buy 500 ETH in one step.

Here, we can make another interesting graphical interpretation. Since the pool has fees, the value of the liquidity parameter $L$ increases after the first purchase (see Equation 2.9). Thus, the points representing the different states of the pool belong to different multiplicative inverse curves. In Figure 2-7, we can see the difference between buying 500 ETH in only one trade and buying 500 ETH in two steps.

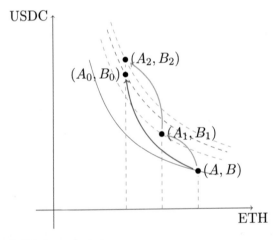

$(A, B)$: Initial pool state.
$(A_0, B_0)$: Pool state after one trade of 500 ETH.
$(A_1, B_1)$: Pool state after one trade of 250 ETH.
$(A_2, B_2)$: Pool state after two consecutive trades of 250 ETH
each.

***Figure 2-7.*** *Comparison between making a trade in one step and dividing it into two trades in a pool with fees*

The following proposition formalizes and generalizes the results of the previous example.

**Proposition 2.1.** *Consider a liquidity pool with two tokens X and Y and with fee $\phi \in [0, 1)$. Let A be the balance of token X in the pool and let B be the balance of token Y in the pool. Let $a_1, a_2 > 0$ such that $a_1 + a_2 < A$. Let $b_0$ be the amount of token Y needed to buy $a_1 + a_2$ tokens X, let $b_1$ be the amount of token Y needed to buy $a_1$ tokens X, and let $b_2$ be the amount of token Y needed to buy $a_2$ tokens X after $a_1$ tokens X have been purchased. Then $b_0 \leq b_1 + b_2$ and the equality holds if and only if $\phi = 0$.*

*Proof.* Let $A' = A - a_1$ and $A'' = A - a_1 - a_2$. From Equation 2.8, we obtain that

$$b_0 = \frac{(a_1 + a_2)B}{(1-\phi)A''}$$

$$b_1 = \frac{a_1 B}{(1-\phi)A'}$$

$$b_2 = \frac{a_2(B+b_1)}{(1-\phi)A''}$$

We have that

$$b_1 + b_2 - b_0 = \frac{a_1 B}{(1-\phi)A'} + \frac{a_2(B+b_1)}{(1-\phi)A''} - \frac{(a_1+a_2)B}{(1-\phi)A''}$$

$$= \frac{1}{(1-\phi)A''}\left( \frac{a_1 BA''}{A'} + a_2(B+b_1) - (a_1+a_2)B \right)$$

$$= \frac{1}{(1-\phi)A''}\left( \frac{a_1 B(A-a_1-a_2)}{A-a_1} + a_2 b_1 - a_1 B \right)$$

$$= \frac{1}{(1-\phi)A''}\left( a_1 B - \frac{a_1 a_2 B}{A-a_1} + a_2 b_1 - a_1 B \right)$$

$$= \frac{1}{(1-\phi)A''}\left( a_2 b_1 - \frac{a_1 a_2 B}{A-a_1} \right)$$

$$= \frac{1}{(1-\phi)A''}\left(\frac{a_1a_2B}{(1-\phi)A'} - \frac{a_1a_2B}{A'}\right)$$

$$= \frac{a_1a_2B}{(1-\phi)A'A''}\left(\frac{1}{(1-\phi)} - 1\right)$$

$$= \frac{a_1a_2B}{(1-\phi)A'A''}\frac{\phi}{(1-\phi)}.$$

Thus, $b_1 + b_2 - b_0 \geq 0$ and $b_1 + b_2 - b_0 = 0$ if and only if $\phi = 0$. The result follows. $\qquad\square$

## 2.2.3 Impact of the Trade Size on the Average Purchase Price

In practice, traders often buy many tokens at once, and as we have seen in Example 2.2, every token costs more than the previous one. We will now analyze this in more detail. Consider a liquidity pool with tokens $X$ and $Y$ and trading fee $\phi$. Let $A$ and $B$ be the balances of tokens $X$ and $Y$ in the pool. Suppose that a trader deposits an amount $b$ of token $Y$ and receives an amount $a$ of token $X$ (hence, they buy token $X$).

Applying Equation 2.8, we obtain that the amount of token $Y$ that the trader has to deposit (pay) per token $X$ that the trader receives is given by the following formula:

$$\frac{b}{a} = \frac{B}{(1-\phi)(A-a)}. \tag{2.10}$$

This can be interpreted as the average purchase price in token $Y$ per unit of token $X$. The total amount of token $Y$ that has to be deposited is given by Equation 2.8:

$$b = \frac{aB}{(1-\phi)(A-a)}.$$

We will study these formulae in the following example.

**Example 2.4** (Purchase price impact). Consider a liquidity pool with B = 40,000,000 USDC and $A$ = 10,000 ETH, having a fee of 0.3%. In Figure 2-8, we plot the price that the trader has to pay for each unit of ETH (in terms of USDC) as a function of the amount of ETH they want to buy (Equation 2.10). Note that it is an increasing function. In Figure 2-8, we also plot the amount of USDC deposited as a function of the received amount $a$ of ETH (Equation 2.8). We observe that it is a convex function.

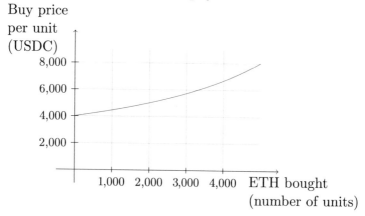

**Price that a trader has to pay for each unit of ETH**

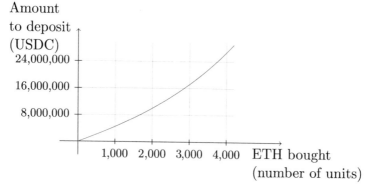

**Total amount to deposit**

**Figure 2-8.** *Price impact: impact of the traded amount on the purchase price*

## 2.2.4 Impact of the Trade Size on the Average Sell Price

We will now make a similar analysis from a seller's perspective. As in the previous subsection, consider a liquidity pool with reserve amounts $A$ of token $X$ and $B$ of token $Y$, and suppose that a trader deposits an amount $b$ of token $Y$ and receives an amount $a$ of token $X$ (and so, the trader sells token $X$). Applying Equation 2.7, we obtain that the amount of token $X$ that the trader receives per token $Y$ that the trader deposits is given by the following formula:

$$\frac{a}{b} = \frac{A(1-\phi)}{B+(1-\phi)b}. \tag{2.11}$$

This can be interpreted as the average sell price in token $X$ per unit of token $Y$. The total amount of token $X$ received is given by Equation 2.7:

$$a = \frac{A(1-\phi)b}{B+(1-\phi)b}.$$

We will study these formulae in the following example.

**Example 2.5** (Sell price impact). Consider a liquidity pool with reserves $A = 40{,}000{,}000$ USDC and $B = 10{,}000$ ETH, having a 0.3% fee. Suppose that a trader sells ETH; that is, the trader deposits an amount $b$ of ETH and receives an amount $a$ of USDC. Using Equation 2.11, we plot in Figure 2-9 the price of each unit of ETH in terms of USDC as a function of the deposited amount $b$ of ETH. Observe that it is a decreasing function. In Figure 2-9, we also plot the amount of USDC received as a function of the deposited amount $b$ of ETH (Equation 2.7). We observe that it is a concave function.

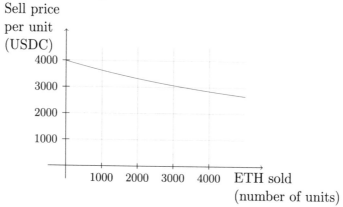

Sell price of each unit of ETH

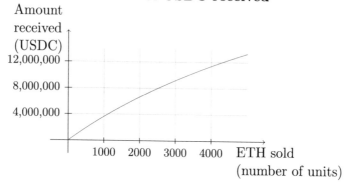

Amount of USDC received

**Figure 2-9.**  *Price impact: impact of the amount traded on the sell price*

## 2.2.5 Impact of the Trade Size on the Price Growth Ratio

We will now study the impact of the trade size on the growth of the effective price paid by a trader. To this end, consider a Uniswap v2 liquidity pool with tokens $X$ and $Y$ and a trading fee of 0.3%. Let $A$ and $B$ be the balances of the tokens $X$ and $Y$, respectively. Note that the current spot price is $p = \dfrac{B}{A}$. We want to compute the growth of the spot price after a trader buys an amount $a$ of token $X$. Note that the amount of token $Y$ the trader has to deposit to buy an amount $a$ of token $X$ is given by Equation 2.8. The updated balances after the trade are

$$A' = A - a,$$

$$B' = B + \frac{aB}{(1-\phi)(A-a)},$$

and hence, the spot price after the trade is

$$p' = \frac{B'}{A'}.$$

Let $r = \dfrac{a}{A}$; that is, $r$ is the fraction of token $X$ that the trader is buying (with respect to the balance of token $X$ in the pool). Using the fact that $p = \dfrac{B}{A}$, we can write the updated spot price $p'$ in terms of $p$ and $r$ as follows:

$$p' = \frac{B + \dfrac{aB}{(1-\phi)(A-a)}}{A-a} = \frac{p + \dfrac{p}{(1-\phi)(r^{-1}-1)}}{1-r}$$

$$= p \cdot \frac{1 + (r^{-1}-1)(1-\phi)}{(1-r)(r^{-1}-1)(1-\phi)}.$$

Hence, the price growth ratio is

$$\frac{p'}{p} = \frac{1 + (r^{-1}-1)(1-\phi)}{(1-r)(r^{-1}-1)(1-\phi)}$$

and it does not depend on the state of the pool. The percentage of increment in the price is given by $\left( \dfrac{p'}{p} - 1 \right) \cdot 100$. In Figure 2-10, we plot (in solid line) the price increment as a function of the fraction $r$ of token $X$ bought (in percentage) in the range (0, 10], and we compare it with a linear function.

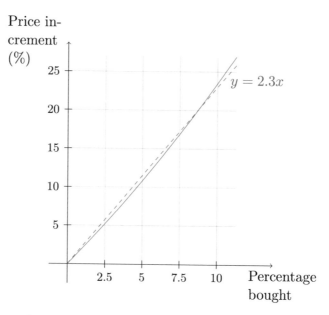

**Figure 2-10.** *Percentage of price increment as a function of the percentage of token X bought*

In Figure 2-11, we plot again (in solid line) the price increment as a function of the fraction $r$ of token $X$ bought, this time in the range (0, 30], and the same linear function as before.

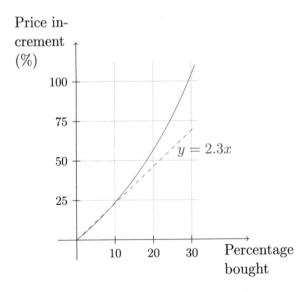

*Figure 2-11.*  *Percentage of price increment as a function of the percentage of token X bought*

# 2.3 Providing Liquidity

In this section, we will explain how liquidity providers can add liquidity to a Uniswap v2 pool and how they can withdraw the liquidity they have deposited. We will also analyze the possible impermanent losses that a liquidity provider might face. In the last part of this section, we will explain how to set a fair price for a liquidity provider's position.

## 2.3.1 Minting LP Tokens

When a liquidity provider wants to provide liquidity to a Uniswap v2 pool, they need to deposit amounts of tokens $X$ and $Y$ that are in a proportion defined by the state of the pool. By doing so, they will be given specific tokens, called *liquidity pool tokens*–or *LP tokens*–which represent the share of the pool they own. The liquidity provider will also earn trading fees according to that share. Note that that share will vary when either new liquidity providers enter the pool or existing ones leave the pool.

We will see now how the process for providing liquidity works. Consider a Uniswap v2 liquidity pool with tokens $X$ and $Y$. Let $A$ and $B$ be the corresponding amounts of tokens $X$ and $Y$ in the pool. If a liquidity provider wants to add liquidity to the pool, they have to deposit an amount $a$ of token $X$ and an amount $b$ of token $Y$ satisfying

$$\frac{b}{a} = \frac{B}{A}, \tag{2.12}$$

that is, the amounts of tokens $X$ and $Y$ that the liquidity provider deposits have to be in the same proportion as that of the pool reserves. Observe that the spot price after the deposit is $\frac{B+b}{A+a}$, which is the same as the original spot price $\frac{B}{A}$ since $\frac{b}{a} = \frac{B}{A}$ (and thus, $A(B+b) = B(A+a)$).

When liquidity providers add liquidity to a pool, they are sent LP tokens in return. The LP tokens act as a receipt of the liquidity providers' provision and represent their share of the liquidity in the pool. The total amount of LP tokens within a pool is dynamic. The AMM keeps track of how many LP tokens it has created and given in return for liquidity. When a new user provides liquidity, the AMM mints an appropriate amount of LP tokens, which is sent to the new liquidity provider, representing the proportion of liquidity that they have provided (see Figure 2-12). In this way, the relative shares of the pool for both previous and new liquidity providers are kept accurate.

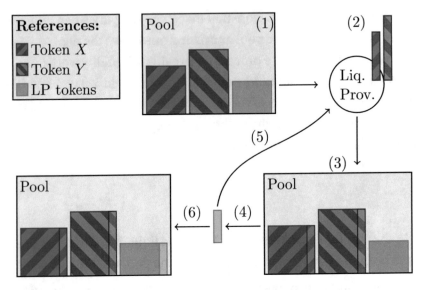

(1) The pool has a certain amount of reserves and tracks the amount of LP tokens in circulation.

(2) The liquidity provider deposits certain amounts of tokens $X$ and $Y$ into the pool. These amounts must be proportional to the balances of tokens $X$ and $Y$ in the pool.

(3) The tokens deposited are added to the pool reserves.

(4) The AMM mints an amount of LP tokens, which is proportional to the amount of tokens deposited.

(5) The newly minted LP tokens are given to the liquidity provider.

(6) The amount of LP tokens in circulation is updated.

***Figure 2-12.***  *Minting LP tokens*

Concretely, consider a Uniswap v2 liquidity pool with tokens $X$ and $Y$, and let $A$ and $B$ be the corresponding amounts of tokens $X$ and $Y$ in the pool. Let $M$ be the amount of existent LP tokens. Suppose that a liquidity provider wants to add liquidity to the pool by depositing an amount $a$ of token $X$ and an amount $b$ of token $Y$. Recall that the amounts $a$ and $b$ must satisfy that

$$\frac{b}{a} = \frac{B}{A}$$

or equivalently,

$$\frac{a}{A} = \frac{b}{B}.$$

Let

$$q = \frac{a}{A}.$$

The new liquidity provider will receive an amount $qM$ of LP tokens. This means that an amount $qM$ of LP tokens will be newly minted and that the total amount of existent LP tokens will be updated to $M + qM = (1 + q)M$.

The Uniswap v2 AMM does not require the liquidity provider to find the exact amounts to deposit. Instead, given amounts $C_X$ of token $X$ and $C_Y$ of token $Y$ that the liquidity provider has available for the deposit, the protocol computes the maximum amounts of tokens $X$ and $Y$ that the liquidity provider can deposit to preserve the pool proportions, and gives the remaining amount of either token $X$ or token $Y$ that was not deposited back to the liquidity provider, together with the LP tokens that correspond to the deposit. This feature is particularly useful due to the ever-changing nonpredictable pool state arising from the fact that the AMM is decentralized. Indeed, if a liquidity provider computes at a certain moment the amounts that they need to deposit, by the time the transaction is processed, the pool state could be different, since other pool transactions could have been processed first, and hence, the amounts needed for the liquidity deposit might have changed.

We will see now how the liquidity deposit works. As before, let $A$ and $B$ be the corresponding amounts of tokens $X$ and $Y$ in the pool and let $M$ be the amount of existent LP tokens. We will consider three cases.

- Case 1: $\dfrac{C_Y}{C_X} = \dfrac{B}{A}$.

In this case, we have that the amounts $C_X$ and $C_Y$ are already in the right proportions, so we define $q = \dfrac{C_X}{A}$. Then, the liquidity provider will receive an amount $qM$ of LP tokens. Note that

$$qM = \frac{C_X}{A} M = \frac{C_Y}{B} M.$$

- Case 2: $\dfrac{C_Y}{C_X} > \dfrac{B}{A}$

In this case, we will find an amount $\Delta_Y > 0$ such that

$$\frac{C_Y - \Delta_Y}{C_X} = \frac{B}{A}.$$

Clearly,

$$\frac{C_Y - \Delta_Y}{C_X} = \frac{B}{A} \Leftrightarrow C_Y - \Delta_Y = \frac{BC_X}{A} \Leftrightarrow \Delta_Y = C_Y - \frac{BC_X}{A}.$$

Since $\dfrac{C_Y}{C_X} > \dfrac{B}{A}$, we obtain that $C_Y > \dfrac{BC_X}{A}$. Then $C_Y - \dfrac{BC_X}{A} > 0$.

Thus, we take $\Delta_Y = C_Y - \dfrac{BC_X}{A}$, and hence, $\dfrac{C_Y - \Delta_Y}{C_X} = \dfrac{B}{A}$. This means

that the amounts $C_X$ and $C_Y - \Delta_Y$ are in the right proportions. We define

then $q = \dfrac{C_X}{A}$. Hence, the liquidity provider will be given back an amount

$\Delta_Y$ of token $Y$ and will receive an amount $qM$ of LP tokens. Note that

$$qM = \frac{C_X}{A}M = \frac{C_Y - \Delta_Y}{B}M < \frac{C_Y}{B}M.$$

- Case 3: $\dfrac{C_Y}{C_X} < \dfrac{B}{A}$

In this case, we will find an amount $\Delta_X > 0$ such that

$$\frac{C_Y}{C_X - \Delta_X} = \frac{B}{A}.$$

Clearly,

$$\frac{C_Y}{C_X - \Delta_X} = \frac{B}{A} \Leftrightarrow C_X - \Delta_X = \frac{AC_Y}{B} \Leftrightarrow \Delta_X = C_X - \frac{AC_Y}{B}.$$

Since $\dfrac{C_Y}{C_X} < \dfrac{B}{A}$, we obtain that $\dfrac{AC_Y}{B} < C_X$. Then $C_X - \dfrac{AC_Y}{B} > 0$.

Thus, we take $\Delta_X = C_X - \dfrac{AC_Y}{B}$ and hence $\dfrac{C_Y}{C_X - \Delta_X} = \dfrac{B}{A}$. This means

that the amounts $C_X - \Delta_X$ and $C_Y$ are in the right proportions. We define

then $q = \dfrac{C_X - \Delta_X}{A}$. Hence, the liquidity provider will be given back an

amount $\Delta_X$ of token $X$ and will receive an amount $qM$ of LP tokens.
Note that

$$qM = \frac{C_Y}{B}M = \frac{C_X - \Delta_X}{A}M < \frac{C_X}{A}M.$$

Summing up, in any case, the liquidity provider will receive an amount
of LP tokens equal to

$$\min\left\{ \frac{C_X}{A}M, \frac{C_Y}{B}M \right\}.$$

In the code given in Listing 2-2, we can find the previous formula.[3] The translation between the variables of the code and our notation is

$$\begin{aligned}
\texttt{totalsupply} &= M, \\
\texttt{amount0} &= C_X, \\
\texttt{amount1} &= C_Y, \\
\texttt{\_eserve0} &= A, \\
\texttt{\_eserve1} &= B, \\
\texttt{liquidity} &= qM.
\end{aligned}$$

***Listing 2-2.*** mint function of the smart contract of Uniswap v2

```
function mint(address to) external lock returns (uint
↪  liquidity) {
    (uint112 _reserve0, uint112 _reserve1,) =
↪  getReserves(); // gas savings
    uint balance0 =
↪  IERC20(token0).balanceOf(address(this));
    uint balance1 =
↪  IERC20(token1).balanceOf(address(this));
    uint amount0 = balance0.sub(_reserve0);
    uint amount1 = balance1.sub(_reserve1);

    bool feeOn = _mintFee(_reserve0, _reserve1);
    uint _totalSupply = totalSupply; // gas savings,
↪  must be defined here since totalSupply can update
↪  in _mintFee
    if (_totalSupply == 0) {
        liquidity =
↪  Math.sqrt(amount0.mul(amount1)).sub(MINIMUM_LIQUIDITY);
        _mint(address(0), MINIMUM_LIQUIDITY); //
```

---

[3] This code can be found in https://github.com/Uniswap/v2-core/blob/master/contracts/UniswapV2Pair.sol

```
↪  permanently lock the first MINIMUM_LIQUIDITY
↪  tokens
     } else {
         liquidity = Math.min(amount0.mul(_totalSupply)
↪  / _reserve0, amount1.mul(_totalSupply) /
↪  _reserve1);
     }
     require(liquidity > 0, 'UniswapV2:
↪  INSUFFICIENT_LIQUIDITY_MINTED');
     _mint(to, liquidity);

     _update(balance0, balance1, _reserve0, _reserve1);
     if (feeOn) kLast = uint(reserve0).mul(reserve1);
↪  // reserve0 and reserve1 are up-to-date
     emit Mint(msg.sender, amount0, amount1);
}
```

## Starting the Pool

Suppose that a liquidity provider starts a liquidity pool adding amounts $a$ and $b$ of tokens $X$ and $Y$, respectively. When starting the pool, the amounts $a$ and $b$ can be in any proportion since there are no balances to compare with. However, Uniswap v2 requires the initial amounts $a$ and $b$ to satisfy that $\sqrt{a \cdot b} > 1{,}000$. In this case, the liquidity provider will receive an amount of LP tokens defined by

$$\text{LP tokens received}\,(\text{start}) = \sqrt{a \cdot b} - 1{,}000.$$

In addition, 1,000 LP tokens will be burned–that is, sent to a special address where the funds can never be retrieved by anyone. It is worth mentioning that this amount of 1,000 is added to the totalSupply variable, as it can be seen from the code given in Listing 2-3. Hence, the totalSupply variable will always be bounded below by 1,000. As it is explained in the

Uniswap v2 documentation,[4] this is done in order to ameliorate rounding errors, and the burned amount will generally represent a negligible value.

***Listing 2-3.*** Portion of the `mint` function of the smart contract of Uniswap v2

```
uint public constant MINIMUM_LIQUIDITY = 10**3;

// ... //

function mint(address to) external lock returns (uint liquidity) {
// ... //

    if (_totalSupply == 0) {
        liquidity =
↳   Math.sqrt(amount0.mul(amount1)).sub(MINIMUM_LIQUIDITY);
        _mint(address(0), MINIMUM_LIQUIDITY); //
↳   permanently lock the first MINIMUM_LIQUIDITY
↳   tokens
    } else {
        liquidity = Math.min(amount0.mul(_totalSupply)
↳   / _reserve0, amount1.mul(_totalSupply) /
↳   _reserve1);
    }
    require(liquidity > 0, 'UniswapV2:
↳   INSUFFICIENT_LIQUIDITY_MINTED');

// ... //

function _mint(address to, uint value) internal {
    totalSupply = totalSupply.add(value);
    balanceOf[to] = balanceOf[to].add(value);
```

---

[4] https://docs.uniswap.org/protocol/V2/concepts/protocol-overview/smart-contracts#minimum-liquidity

```
    emit Transfer(address(0), to, value);
}
```

## 2.3.2 Burning LP Tokens

Liquidity providers can also remove liquidity from the pool. In this case, they have to send their LP tokens to the pool, and they will be given certain amounts of the pool tokens according to their share of the pool, as Figure 2-13.

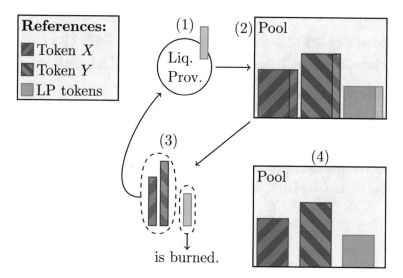

(1) The liquidity provider wants to redeem his/her share of the pool, which is represented by the amount of LP tokens he/she holds.

(2) The AMM computes the share owned by the liquidity provider and the corresponding amounts of tokens $X$ and $Y$ that he/she owns according to that share.

(3) The AMM gives those amounts of tokens $X$ and $Y$ to the liquidity provider and burns the LP tokens received from him/her.

(4) The balances of each token in the pool and the amount of LP tokens in circulation are updated.

***Figure 2-13.*** *Burning LP tokens*

We will now explain in detail how this works. Consider a Uniswap v2 liquidity pool with tokens $X$ and $Y$. Let $A$ and $B$ be the corresponding amounts of tokens $X$ and $Y$ in the pool. Let $M$ be the amount of existent LP tokens and let $m$ be the amount of LP tokens that the liquidity provider wants to redeem. Clearly, the amount $m$ of LP tokens corresponds to a share $\dfrac{m}{M}$ of the pool. Therefore, in exchange for the amount $m$ of LP tokens, the liquidity provider will receive an amount $\dfrac{m}{M}A$ of token $X$ and an amount $\dfrac{m}{M}B$ of token $Y$. The amount $m$ of LP tokens deposited by the liquidity provider will be burned, and the amount of existent LP tokens will be updated to $M - m$. Clearly, the pool reserves will also be updated to $A - \dfrac{m}{M}A$ for token $X$ and $B - \dfrac{m}{M}B$ for token $Y$.

The formula for the amount of each token that the liquidity provider has to receive when redeeming LP tokens is covered by the burn function of Uniswap v2,[5] which is given in Listing 2-4, with the following translation between the variables of the code and our notation:

$$\_\texttt{totalsupply} = M,$$
$$\_\texttt{balance0} = A,$$
$$\_\texttt{balance1} = B,$$
$$\_\texttt{liquidity} = m,$$

**Listing 2-4.** burn function of the smart contract of Uniswap v2

```
// this low-level function should be called from a
↪  contract which performs important safety checks
function burn(address to) external lock returns (uint
↪  amount0, uint amount1) {
    (uint112 _reserve0, uint112 _reserve1,) =
↪  getReserves(); // gas savings
    address _token0 = token0;
```

---

[5] https://github.com/Uniswap/v2-core/blob/master/contracts/UniswapV2Pair.sol

```
↪  // gas savings
   address _token1 = token1;
↪  // gas savings
   uint balance0 =
↪  IERC20(_token0).balanceOf(address(this));
   uint balance1 =
↪  IERC20(_token1).balanceOf(address(this));
   uint liquidity = balanceOf[address(this)];

   bool feeOn = _mintFee(_reserve0, _reserve1);
   uint _totalSupply = totalSupply; // gas savings,
↪  must be defined here since totalSupply can update
↪  in _mintFee
   amount0 = liquidity.mul(balance0) / _totalSupply;
↪  // using balances ensures pro-rata distribution
   amount1 = liquidity.mul(balance1) / _totalSupply;
↪  // using balances ensures pro-rata distribution
   require(amount0 > 0 && amount1 > 0, 'UniswapV2:
↪  INSUFFICIENT_LIQUIDITY_BURNED');
   _burn(address(this), liquidity);
   _safeTransfer(_token0, to, amount0);
   _safeTransfer(_token1, to, amount1);
   balance0 =
↪  IERC20(_token0).balanceOf(address(this));
   balance1 =
↪  IERC20(_token1).balanceOf(address(this));

   _update(balance0, balance1, _reserve0, _reserve1);
   if (feeOn) kLast = uint(reserve0).mul(reserve1);
↪  // reserve0 and reserve1 are up-to-date
   emit Burn(msg.sender, amount0, amount1, to);
}
```

## Distribution of Fees

We explained before how traders are charged a fee on the way in, meaning that a certain amount of the token they deposit is kept as a fee and the remaining amount is actually traded (see Figure 2-3). That fee is not shared among the liquidity providers immediately, since doing that would be not only impractical but also very expensive in terms of gas fees. Instead, the fee is added to the pool so that the whole value of the pool increases and thus the value of each liquidity provider's share increases as well. In this way, when the liquidity provider decides to redeem their LP tokens, they obtain an added value from the fees that have accumulated in the pool (see Figure 2-14).

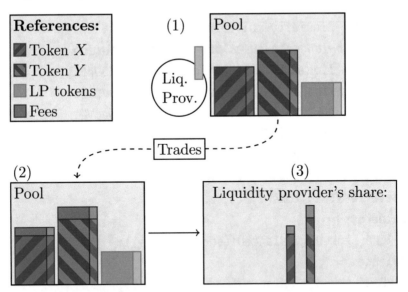

(1) A liquidity provider owns a share of the pool.
(2) As trades are performed, trading fees accrue in the pool.
(3) Now, the liquidity provider also owns a portion of the collected fees.

*Figure 2-14. Visualizing how liquidity providers earn trading fees*

# 2.3.3 Pool Value and Impermanent Loss

Consider a Uniswap v2 liquidity pool with tokens $X$ and $Y$. Let $A$ and $B$ be the balances of tokens $X$ and $Y$, respectively. From Equation 2.4, we know that the spot price of token $X$ in terms of token $Y$ is $\dfrac{B}{A}$, and hence, the value of the amount $A$ of token $X$ that the pool has in terms of token $Y$ is $\dfrac{B}{A} \cdot A$. And since the value of the amount $B$ of token $Y$ in the pool is $B$, we obtain that the total value of the pool in terms of token $Y$ is

$$\frac{B}{A} \cdot A + B = 2B.$$

Suppose that a liquidity provider owns a share $\rho$ of all the existent LP tokens. If they decide to redeem their LP tokens, they will receive an amount $\rho A$ of token $X$ and an amount $\rho B$ of token $Y$. Therefore, the value of their position in terms of token $Y$ is

$$\frac{B}{A} \cdot \rho A + \rho B = \rho \left( \frac{B}{A} \cdot A + B \right) = \rho \cdot 2B,$$

that is, a fraction $\rho$ of the whole pool value, as expected.

## Computing Impermanent Loss Without Fees

Consider a Uniswap v2 liquidity pool with no fees having 9 ETH and 36,000 USDC. Suppose that the existent amount of LP tokens is 90 and that a liquidity provider adds 1 ETH and 4,000 USDC to the pool. Thus, the pool balances after the deposit are 10 ETH and 40,000 USDC, and the liquidity parameter $L$ satisfies $L^2 = 400,000$. Note that the liquidity provider earns an amount of $\dfrac{1}{9} \cdot 90 = 10\,\text{LP}$ tokens and that the total amount of existent LP tokens is now 100. Hence, the new liquidity provider owns a share of 10% of the existent LP tokens and thus of the total pool liquidity. Note that the spot price of ETH before and after the deposit is 4,000 USDC/ETH.

Now, suppose that the spot price for 1 ETH goes up to 6,000 USDC. The updated amounts $A'$ and $B'$ of ETH and USDC can be computed from the following equations:

$$A' \cdot B' = L^2 = 400{,}000$$

$$\frac{B'}{A'} = 6{,}000$$

We obtain that

$$A' = \sqrt{\frac{400{,}000}{6{,}000}} \approx 8.1649658$$

and

$$B' = \sqrt{400{,}000 \cdot 6{,}000} \approx 48{,}989.79.$$

That is, the liquidity pool has now approximately 8.1649658 ETH and 48,989.79 USDC. If the liquidity provider withdraws their position, they will receive 0.81649658 ETH and 4,898.979 USDC since they own a share of 10% of the existent LP tokens. This results in a new portfolio value of approximately

$$0.81649658 \cdot 6{,}000 + 4{,}898.979 \approx 9{,}797.96 \ \ \text{USDC}.$$

However, if they had never decided to be a liquidity provider, their original position would have been worth

$$1 \cdot 6{,}000 + 4{,}000 = 10{,}000 \ \ \text{USDC}.$$

Therefore, they are facing a loss of approximately 202.04 USDC, which amounts to a loss of approximately 2.02% with respect to holding the assets. This loss is called *impermanent loss* (*or divergence loss*) and is due to the fact that the amounts of the tokens are moving on a multiplicative inverse curve. It is called impermanent because if the ETH price reverts to the original 4,000 USDC/ETH (which is the spot price at the moment of the deposit), then the pool balances will be again 10 ETH and 40,000 USDC, and in consequence, the liquidity provider's position will be the same as the one at the beginning and there will be no loss. But if the liquidity provider decides to withdraw their position when the price is 6,000 USDC/ETH, then the previous loss becomes permanent.

We will now generalize what we have seen previously. Consider a Uniswap v2 liquidity pool (with no fees) having two tokens $X$ and $Y$ with balances $A_0$ and $B_0$, respectively. Let $M$ be the amount of existent LP tokens. Suppose that a liquidity provider adds an amount $a$ of token $X$ and an amount $b$ of token $Y$ satisfying that $\dfrac{a}{A_0} = \dfrac{b}{B_0}$. Let $q = \dfrac{a}{A_0}$. We know that the liquidity provider receives an amount $qM$ of LP tokens. In addition, after the deposit, the pool reserves are $A = A_0 + a$ for token $X$ and $B = B_0 + b$ for token $Y$, and the total amount of LP tokens is $(1 + q)M$. Note that the liquidity provider owns a share $\rho = \dfrac{qM}{(1+q)M} = \dfrac{q}{1+q}$ of the pool. Since $q = \dfrac{a}{A_0} = \dfrac{b}{B_0}$, we obtain that

$$\rho = \frac{q}{1+q} = \frac{\dfrac{b}{B_0}}{1+\dfrac{b}{B_0}} = \frac{b}{B_0 + b} = \frac{b}{B}.$$

Let $p = \dfrac{B}{A}$ be the spot price after the deposit (which is the same as the spot price before the deposit, as we have previously proved). Hence,

$$p = \frac{B}{A} = \frac{B_0}{A_0} = \frac{b}{a}.$$

Suppose now that, after some trading activities, the price shifts from $p$ to $p'$. Let $A'$ and $B'$ be the balances of tokens $X$ and $Y$, respectively, after the price change. Hence, $p' = \dfrac{B'}{A'}$. And since the pool has no fees, we have that $A'B' = AB$. Thus,

$$B' = p'A' = p'\frac{AB}{B'}.$$

Hence,

$$\left(B'\right)^2 = p'AB = p'\frac{A}{B}B^2 = \frac{p'}{p}B^2.$$

Thus,

$$B' = \sqrt{\frac{p'}{p}}B.$$

From the analysis at the beginning of this subsection, we know that the value of the liquidity provider's position in terms of token $Y$ is $2\rho B'$. Note that

$$2\rho B' = 2\rho B\sqrt{\frac{p'}{p}} = 2b\sqrt{\frac{p'}{p}}.$$

If the liquidity provider had held the amounts of both tokens instead of depositing them into the pool, the value of their position in terms of token $Y$ would have been

$$b + p'a = b + p'\frac{b}{p} = b\left(1 + \frac{p'}{p}\right).$$

Hence, the impermanent loss in terms of token $Y$ is

$$2b\sqrt{\frac{p'}{p}} - b\left(1 + \frac{p'}{p}\right),$$

and therefore, the fraction of impermanent loss in terms of token $Y$ with respect to the value of the original position is

$$\frac{2b\sqrt{\dfrac{p'}{p}} - b\left(1 + \dfrac{p'}{p}\right)}{b\left(1 + \dfrac{p'}{p}\right)} = \frac{2b\sqrt{\dfrac{p'}{p}}}{b\left(1 + \dfrac{p'}{p}\right)} - 1 = \frac{2\sqrt{\dfrac{p'}{p}}}{1 + \dfrac{p'}{p}} - 1.$$

We plot in Figure 2-15 the percentage of impermanent loss as a function of the spot price (expressed as a percentage of the price at the moment of the deposit).

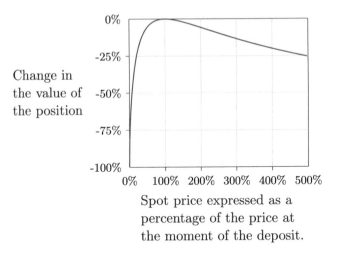

**Figure 2-15.** *Losses of liquidity providers due to price variation compared to holding the funds supplied*

## Accounting for Fees

The previous analysis did not take fees into consideration. In a pool with a nonzero trading fee, all trading activities add liquidity to the pool in the form of collected fees, which increase the value of the liquidity providers' positions and hence reduce their impermanent losses.

In order to understand how this works, we will analyze the following example. Consider a liquidity pool with ETH and USDC and with fee $\phi = 0.003$. Assume that the pool has 1,000 ETH and 4,000,000 USDC. Note that the spot price is $p = 4,000$ USDC/ETH. Suppose that a liquidity provider owns a share of 10% of the pool.

Assume that a trader buys 1 ETH in the pool. Applying Equation 2.8, we obtain that the amount of USDC that the trader has to deposit is

$$b = \frac{1 \cdot 4,000,000}{(1 - 0.003) \cdot 999} \approx 4,016.$$

Hence, the updated balances of the pool are

$$A' = 999 \ \text{ETH},$$
$$B' \approx 4{,}004{,}016 \ \text{USDC},$$

and the new spot price is $p' \approx 4{,}008.024$ USDC/ETH.

Observe that the amount of the fee is included in $B'$ since the trader was charged the fee on their deposit. Suppose that no other trades are performed and no further liquidity is added. Then, the value of the whole pool is $B' + p' \cdot A' (= 2B')$. And since the liquidity provider owns a share of 10% of the pool, the value of their position is approximately

$$800{,}803.2 \ \text{USDC}.$$

If the liquidity provider had not deposited their tokens into the pool, the value in USDC of their position would have been approximately

$$100 \cdot 4{,}008.024 + 400{,}000 = 800{,}802.4$$

Note that in this case, the impermanent loss is a gain due to the collected fees:

$$800{,}803.2 - 800{,}802.4 = 0.8 \ \text{USDC}$$

In the case of a liquidity pool without trading fees ($\phi = 0$) and considering the same situation as the previous one, we could perform similar computations to obtain that the liquidity provider would face an impermanent loss of approximately 0.4 USDC.

For liquidity pools with massive trading activity around the starting price, the trading fee becomes more significant for the value of the liquidity provider's position, and the impermanent loss can sometimes be compensated, resulting in gains for the liquidity providers.

## 2.3.4  LP Token Swap

In this subsection, we will study the problem of swapping LP tokens of two different liquidity pools. To calculate the fair swap price, we will need to compute the values of those LP tokens.

Consider two liquidity providers, Alice and Bob, who are members of two different liquidity pools $P_1$ and $P_2$, which have liquidity pool tokens $LP_1$ and $LP_2$, respectively. Both Alice and Bob own a certain amount of LP tokens of their respective pools. Let $N_1$ and $N_2$ be the the total minted liquidity tokens in pools $P_1$ and $P_2$, respectively.

We will divide our analysis into two different cases: when the liquidity pools $P_1$ and $P_2$ have at least one common token and when they do not.

### Common Token Case

Suppose first that the liquidity pools $P_1$ and $P_2$ have (at least) one common token $Y$. We want to compute the fair swap price between Alice's and Bob's LP tokens. In other words, we want to calculate how the values of the tokens $LP_1$ and $LP_2$ are related.

Let $B_1$ and $B_2$ be the amounts of token $Y$ in pools $P_1$ and $P_2$, respectively. Hence, for $j \in \{1, 2\}$, the total value of pool $P_j$ (in terms of token $Y$) is $2B_j$ and thus, the value of each $LP_j$ token (in terms of token $Y$) is

$$V_{LP_j}(Y) = \frac{2B_j}{N_j}$$

Let $P$ be the the price of one $LP_1$ token in terms of $LP_2$ tokens. That is, one $LP_1$ token is equivalent to $P$ tokens $LP_2$. Hence, one $LP_1$ token and $P$ tokens $LP_2$ have the same value in terms of $Y$ tokens. Thus, $V_{LP_1}(Y) = P \cdot V_{LP_2}(Y)$, and therefore,

$$P = \frac{V_{LP_1}(Y)}{V_{LP_2}(Y)} = \frac{N_2}{N_1} \cdot \frac{B_1}{B_2}. \tag{2.13}$$

Summing up, this means that if Alice gives one LP token from pool $P_1$ to Bob, Alice will get $\dfrac{N_2}{N_1} \cdot \dfrac{B_1}{B_2}$ tokens from pool $P_2$ in return.

**Example 2.6.** Suppose Alice and Bob are liquidity providers with the following data:

|  | Alice (Pool $P_1$) | Bob (Pool $P_2$) |
|---|---|---|
| Total LP tokens | 12,000 | 10,000 |
| LP tokens owned | 100 | 100 |
| Token $X$ | ETH | ETH |
| Token $Y$ | USDC | USDC |
| Amount of token $X$ | 10,000 | 9,750 |
| Amount of token $Y$ | 40,000,000 | 39,000,000 |

Now we apply Equation 2.13 to find the fair swap price between the corresponding LP tokens:

$$P = \frac{10,000}{12,000} \cdot \frac{40,000,000}{39,000,000} \approx 0.8547.$$

Hence, if Alice swaps all her 100 LP tokens from pool $P_1$, she will receive approximately 85.47 LP tokens from pool $P_2$ in return. Note that, although the liquidity of pool $P_2$ is less than that of pool $P_1$, the amount of LP tokens minted in pool $P_2$ is less than that of $P_1$, resulting in a relatively higher price of the token of pool $P_2$ compared with the token of pool $P_1$.

# No In-Common Token Case: Oracle Swap

Suppose now that pools $P_1$ and $P_2$ do not have a common token. In that case, we will use external price data to express amounts of certain tokens $Y_1 \neq Y_2$ in terms of amounts of a chosen token $Z$. We will need the corresponding exchange rates (or prices) to do so.

Let $Y_1$ be a token of pool $P_1$ and let $Y_2$ be a token of pool $P_2$. Suppose that $Y_1 \neq Y_2$. For $j \in \{1, 2\}$, let $B_j$ be the amount of token $Y_j$ in pool $P_j$. Let $Z$ be a token, and for $j \in \{1, 2\}$, let $p_j$ be the price of token $Y_j$ in terms of token $Z$.

Proceeding as in the previous case, we obtain that for $j \in \{1, 2\}$, the value of each $\text{LP}_j$ token in terms of token $Y_j$ is

$$V_{\text{LP}_j}(Y_j) = \frac{2B_j}{N_j}$$

and hence, the value of each $\text{LP}_j$ token in terms of token $Z$ is

$$V_{\text{LP}_j}(Z) = p_j \cdot \frac{2B_j}{N_j}.$$

Let $P$ be the price of one $\text{LP}_1$ token in terms of $\text{LP}_2$ tokens. Hence, $V_{\text{LP}_1}(Z) = P \cdot V_{\text{LP}_2}(Z)$, and thus,

$$P = \frac{V_{\text{LP}_1}(Z)}{V_{\text{LP}_2}(Z)} = \frac{N_2}{N_1} \cdot \frac{B_1}{B_2} \cdot \frac{p_1}{p_2}. \tag{2.14}$$

Summing up, this means that if Alice gives one LP token from pool $P_1$ to Bob, Alice will get $\dfrac{N_2}{N_1} \cdot \dfrac{B_1}{B_2} \cdot \dfrac{p_1}{p_2}$ tokens from pool $P_2$ in return.

**Example 2.7.** Consider two liquidity pools described by the data given in the following table:

|  | **Pool $P_1$** | **Pool $P_2$** |
|---|---|---|
| Total LP tokens | 12,000 | 10,000 |
| Token $X$ | WBTC | XRP |
| Token $Y$ | ETH | DAI |

|                        | Pool $P_1$ | Pool $P_2$ |
|------------------------|------------|------------|
| Amount of token $X$    | 100        | 1,000,000  |
| Amount of token $Y$    | 1,200      | 800,000    |

We will use USDC as the token $Z$ in which prices will be expressed, with exchange rates of 4,000 USDC/ETH and 1 USDC/DAI.

Applying Equation 2.14, we obtain that the fair swap price between the corresponding LP tokens is

$$P = \frac{10,000}{12,000} \cdot \frac{1,200}{800,000} \cdot \frac{4,000}{1} = 5.$$

Thus, each $LP_1$ token is worth $5\,LP_2$ tokens. Note that although pool $P_2$ has minted fewer LP tokens, the total value of pool $P_1$ $(2 \cdot 1,200 \cdot 4,000 = 9,600,000$ USDC$)$ is much higher than that of pool $P_2$($2 \cdot 800,000 \cdot 1 = 1,600,000$ USDC$)$, resulting in a high exchange rate between the corresponding LP tokens.

# 2.4 Motivating DEX Aggregators

A DEX aggregator is a blockchain-based service that functions as an explorer for the prices and liquidity offered by different Decentralized Exchanges and helps traders find the best price for their trades. In addition, DEX aggregators are equipped with an algorithm that allows users to split the trade they want to do in several trades against different AMMs so as to obtain the best possible price for that trade. Clearly, this is a very enticing service. We will develop a simple example to show why this makes sense and how one can split a trade between two different pools to obtain a better price.

Suppose that we want to trade an amount $T$ of ETH among two different ETH/USDC pools. For example, these can be Uniswap v2 pools on two different networks or one Uniswap and one Sushiswap pool on the same network. We want to find the amount of ETH we should trade in each pool in order to receive the maximum possible amount of USDC.

For $j \in \{1, 2\}$, let $A_j$ and $B_j$ be the reserves of pool $J$ of ETH and USDC, respectively, and let $\phi_1$ and $\phi_2$ be the fees of pools 1 and 2, respectively. Note that if we trade an amount $x$ of ETH in the first pool and the remaining amount $T - x$ of ETH in the second pool, from Equation 2.7, we obtain that the amount of USDC received from the first pool is

$$\frac{B_1(1-\phi_1)x}{A_1+(1-\phi_1)x}$$

and the amount of USDC received from the second pool is

$$\frac{B_2(1-\phi_2)(T-x)}{A_2+(1-\phi_2)(T-x)}.$$

In order to maximize the amount of USDC received when trading the amount $T$ of ETH among the two pools, we have to find the maximum of the function $f\colon [0, T] \to \mathbb{R}$ defined by

$$f(x)=\frac{B_1(1-\phi_1)x}{A_1+(1-\phi_1)x}+\frac{B_2(1-\phi_2)(T-x)}{A_2+(1-\phi_2)(T-x)}.$$

For simplicity, let $\overline{\phi_1}=1-\phi_1$ and $\overline{\phi_2}=1-\phi_2$. Note that

$$f(x)=\frac{B_1\overline{\phi_1}x}{A_1+\overline{\phi_1}x}+\frac{B_2\overline{\phi_2}(T-x)}{A_2+\overline{\phi_2}(T-x)}.$$

$$= B_1 \cdot \frac{\overline{\phi_1} x + A_1 - A_1}{A_1 + \overline{\phi_1} x} + B_2 \cdot \frac{\overline{\phi_2}(T - x) + A_2 - A_2}{A_2 + \overline{\phi_2}(T - x)}$$

$$= B_1 \left( 1 - \frac{A_1}{A_1 + \overline{\phi_1} x} \right) + B_2 \left( 1 - \frac{A_2}{A_2 + \overline{\phi_2}(T - x)} \right)$$

$$= B_1 - \frac{A_1 B_1}{A_1 + \overline{\phi_1} x} + B_2 - \frac{A_2}{A_2 + \overline{\phi_2}(T - x)}.$$

Thus, the derivative of $f$ is given by

$$f'(x) = \frac{A_1 B_1 \overline{\phi_1}}{\left(A_1 + \overline{\phi_1} x\right)^2} - \frac{A_2 B_2 \overline{\phi_2}}{\left(A_2 + \overline{\phi_2}(T - x)\right)^2}$$

$$= \frac{A_1 B_1 \overline{\phi_1} \left(A_2 + \overline{\phi_2}(T - x)\right)^2 - A_2 B_2 \overline{\phi_2} \left(A_1 + \overline{\phi_1} x\right)^2}{\left(A_1 + \overline{\phi_1} x\right)^2 \left(A_2 + \overline{\phi_2}(T - x)\right)^2}.$$

Let $g(x)$ denote the numerator of the previous expression, that is,

$$g(x) = A_1 B_1 \overline{\phi_1} \left(A_2 + \overline{\phi_2}(T - x)\right)^2 - A_2 B_2 \overline{\phi_2} \left(A_1 + \overline{\phi_1} x\right)^2.$$

Note that $g(x)$ is just a polynomial of degree 2 and that for any $x_0 \in [0, T]$, $f'(x_0) = 0$ if and only if $g(x_0) = 0$. Moreover, since the denominator of the last expression of $f'(x)$ is always positive, we obtain that for any $x_0 \in [0, T]$, $f'(x_0) > 0$ if and only if $g(x_0) > 0$ (and $f'(x_0) < 0$ if and only if $g(x_0) < 0$). Thus, in order to study the zeros and the sign of $f'$, we can study the zeros and the sign of $g$. Recall that since $g$ is a polynomial of degree 2, the zeros of $g$ can be computed from the quadratic formula. Then the positive and negative regions of $g$ can easily be obtained observing the concavity of $g$ from its leading coefficient.

In the following example, we will show how the previous argument can be applied.

**Example 2.8.** Consider two ETH/USDC liquidity pools with the following reserves:

|      | Pool 1     | Pool 2     |
|------|------------|------------|
| ETH  | 10,000     | 10,300     |
| USDC | 40,000,000 | 39,500,000 |
| $\Phi$ | 0.3%     | 0.3%       |

Suppose that we want to trade a total amount of 1,500 ETH between these two pools. To maximize the amount of USDC received, we have to find the maximum value of the function $f: [0, 1500] \to \mathbb{R}$ defined by

$$f(x) = \frac{40,000,000 \cdot 0.997 \cdot x}{10,000 + 0.997 \cdot x} + \frac{39,500,000 \cdot 0.997 \cdot (1,500 - x)}{10,300 + 0.997 \cdot (1,500 - x)}.$$

We will thus find the zeros of the derivative $f'$ of $f$ and the positive and negative regions of $f'$ in order to obtain the intervals in which $f$ is an increasing function and the intervals in which $f$ is a decreasing function. By the argument preceding this example, it is enough to find the zeros and the positive and negative intervals of the function $g: \mathbb{R} \to \mathbb{R}$ defined by

$$g(x) = A_1 B_1 \overline{\phi_1} \left( A_2 + \overline{\phi_2}(T - x) \right)^2 - A_2 B_2 \overline{\phi_2} \left( A_1 + \overline{\phi_1}x \right)^2,$$

in this case,

$$g(x) = \ 10,000 \cdot 40,000,000 \cdot 0.997 \cdot \left(10,300 + 0.997 \cdot (1,500 - x)\right)^2 - $$
$$10,300 \cdot 39,500,000 \cdot 0.997 \cdot \left(10,000 + 0.997x\right)^2.$$

We can expand the previous expression to get

$$g(x) = -6{,}788{,}534{,}765.05 \cdot x^2 - 17{,}468{,}117{,}760{,}600{,}000 \cdot x + 14{,}923{,}622{,}515{,}700{,}000{,}000.$$

Applying the quadratic formula, we obtain that the roots of $g$ are

$$x_1 \approx -2{,}574{,}033.44 \text{ and } x_2 \approx 854.0514.$$

Since the leading coefficient of $g$ is negative, we obtain that $g$ is positive on the interval $(x_1, x_2)$ and negative on the intervals $(-\infty, x_1)$ and $(x_2, +\infty)$ (and we know that the same holds for $f'$).

And since we are interested in computing the maximum of $f$ on the interval $[0, T]$, from the sign of $f'$, we obtain that $f$ is strictly increasing on the interval $[0, x_2]$ and strictly decreasing on the interval $[x_2, 1{,}500]$. Thus, the function $f$ attains a maximum on $x_2 \approx 854.0514$. Therefore, in order to maximize the amount of USDC received when trading 1,500 ETH, we have to sell 854.0514 ETH on pool 1 and $1{,}500 - 854.0514 = 645.9486$ ETH on pool 2. By doing so, we will obtain $f(854.0514) \approx 5{,}463{,}115.24$ USDC.

Observe that if we had performed the whole trade on pool 1, we would have received $f(1{,}500) \approx 5{,}203{,}775.39$ USDC, and if we had performed the whole trade on pool 2, we would have received $f(0) \approx 5{,}008{,}032.72$ USDC.

As we can see from the previous example, we can obtain a much better price if we split the trade between the two pools in a suitable way. However, in order to do so, we need to have all the information about the state of the two pools, and we also need to perform a lot of mathematical computations. DEX aggregators do all this work for us, and hence, they are a valuable and much appreciated tool.

# 2.5 Summary

In this chapter, we performed a thorough analysis of the Uniswap v2 AMM, introducing the concepts of spot price, effective price, price impact, and impermanent loss. We show how trades are performed in Uniswap v2 and how liquidity can be deposited and withdrawn.

In the next chapter, we will analyze the Balancer AMM, which is a generalization of the Uniswap v2 AMM and allows liquidity pools of more than two assets.

# CHAPTER 3

# Balancer

Balancer generalizes the constant product formula of Uniswap v2, introducing a weighted geometric mean formula that allows the creation of pools that may have more than two assets. In addition, when a Balancer pool is created, each asset of the pool is endowed with a certain weight representing the share of that asset in the pool. This gives additional flexibility to the composition of the pools.

In this chapter, we will describe in detail how the Balancer AMM works. We will first show how the spot price formula and the trading formulae are obtained, and then we will explain how liquidity providers can deposit and withdraw liquidity. Unlike Uniswap, Balancer allows single-asset and all-asset deposits and withdrawals, and we will explain how the four cases are handled. We will also compare the single-asset deposit case with the all-asset deposit case, performing a complete analysis that includes several novel results. It is important to mention that this analysis cannot be found elsewhere.

## 3.1 The Constant Value Function

The Balancer AMM is based on *Balancer's value function*, which is the function $V : (\mathbb{R}_{>0})^N \to \mathbb{R}_{>0}$ that is defined by

$$V(b_1, b_2, \ldots, b_N) = \prod_{j=1}^{N} b_j^{W_j}$$

M. Ottina et al., *Automated Market Makers*, https://doi.org/10.1007/978-1-4842-8616-6_3

where $N$ is the number of different tokens in the pool and for each $j \in \{1, 2, ..., N\}$, the (constant) number $W_j$ is the normalized weight of token $j$ in the pool and the variable $b_j$ represents the balance of token $j$ in the pool. Since $W_1, W_2, ... W_N$ are normalized weights, we require that $W_j \in (0, 1)$ for all $j \in \{1, 2, ..., N\}$ and $\displaystyle\sum_{j=1}^{N} W_j = 1$.

The basic idea of how the Balancer AMM works is similar to the Uniswap v2 AMM: the spot price formula and the trading formulae for a Balancer pool are obtained by setting Balancer's value function equal to a constant value, in a similar way as the corresponding formulae for Uniswap v2 are derived from the constant product formula. Consequently, we obtain that an equally weighted Balancer pool of two tokens is equivalent to a Uniswap v2 pool.

Observe that if $V_0 > 0$, the equation $V(b_1, b_2, ..., b_N) = V_0$ defines an $(N - 1)$-dimensional (differentiable) surface. We can see how this surface looks like in Figure 3-1.

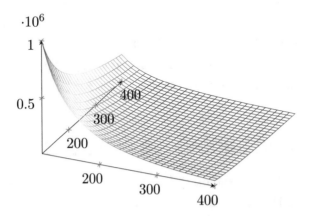

**Figure 3-1.** *Example of a possible two-dimensional surface related to a Balancer's pool*

Note also that for all $i \in \{1, 2, ..., N\}$ and for any $V_0 \in \mathbb{R}$,

$$\frac{\partial\left(V - V_0\right)}{\partial b_i} = \frac{\partial V}{\partial b_i} = W_i b_i^{W_i - 1} \prod_{\substack{j=1 \\ j \neq i}}^{N} b_j^{W_j} \neq 0$$

Hence, when the value function is constrained to a constant value $V_0$ (which belongs to the range of the map $V$), applying the Implicit Function Theorem to the map $V - V_0$, we obtain that any variable $b_i$ can be described as a function of the other variables $b_j, j \neq i$, in a certain neighborhood of any point that belongs to $V^{-1}(V_0)$.

# 3.2 Spot Price

Consider a Balancer pool with $N$ tokens. In order to define the spot price, we will work in a pool with no fees. Let $i, o \in \{1, 2, ..., N\}$. Note that the *effective price* of token $o$ in terms of token $i$ that a trader pays in a certain trade is given by $EP_{o,i} = \dfrac{A_i}{A_o}$, where $A_i$ is the amount of token $i$ being deposited by the trader (and hence that goes into the pool) and $A_o$ is the amount of token $o$ being bought by the trader (and hence that goes out of the pool). The spot price of token $o$ in terms of token $i$ is defined as $SP_{o,i} = \lim_{A_o, A_i \to 0} EP_{o,i}$.

Suppose that a trader wants to buy token $o$ in exchange for token $i$. As in Uniswap v2, in any trade, Balancer's value function is constrained to a constant value $V_0$, and hence, as we have seen before, the variable $b_i$ can be described as a function of the other variables $b_j, j \neq i$.

For each $j \in \{1, 2, ..., N\}$, let $B_j$ be the balance of token $j$ before the trade and let $B_j'$ be the balance of token $j$ after the trade. Let $A_o$ be the amount of token $o$ that the trader receives and let $A_i$ be the amount of token $i$ the trader deposits. Clearly, $B_o' = B_o - A_o$, $B_i' = B_i + A_i$, and $B_j' = B_j$ for all $j \in \{1, 2, ..., N\} - \{i, o\}$. Thus,

$$SP_{o,i} = \lim_{A_o, A_i \to 0} EP_{o,i} = \lim_{A_o, A_i \to 0} \frac{A_i}{A_o} = \lim_{A_o, A_i \to 0} \frac{B_i' - B_i}{B_o - B_o'}$$

$$= -\lim_{A_i \to 0} \frac{B_i' - B_i}{B_o' - B_o} = -\frac{\partial b_i}{\partial b_o}(B_1, B_2, \ldots, B_N)$$

Let $\mathbf{B} = (B_1, B_2, \ldots, B_N)$. Applying the Implicit Function Theorem (to the map $V - V_0$), we obtain that

$$\frac{\partial b_i}{\partial b_o}(\mathbf{B}) = -\frac{\dfrac{\partial V}{\partial b_o}(\mathbf{B})}{\dfrac{\partial V}{\partial b_i}(\mathbf{B})} = -\frac{W_o B_o^{W_o - 1} \displaystyle\prod_{\substack{j=1 \\ j \neq o}}^{N} B_j^{W_j}}{W_i B_i^{W_i - 1} \displaystyle\prod_{\substack{j=1 \\ j \neq i}}^{N} B_j^{W_j}}$$

$$= -\frac{W_o B_o^{W_o - 1} B_i^{W_i} \displaystyle\prod_{\substack{j=1 \\ j \neq o,i}}^{N} B_j^{W_j}}{W_i B_i^{W_i - 1} B_o^{W_o} \displaystyle\prod_{\substack{j=1 \\ j \neq o,i}}^{N} B_j^{W_j}} = -\frac{W_o B_i}{W_i B_o} = -\frac{\dfrac{B_i}{W_i}}{\dfrac{B_o}{W_o}}.$$

Therefore, the formula for the spot price calculation is

$$SP_{o,i} = \frac{\dfrac{B_i}{W_i}}{\dfrac{B_o}{W_o}}. \tag{3.1}$$

In Balancer, the fees are charged *on the way in*. That means that when someone wants to perform a trade depositing a certain amount $a$ of a token $X$ into the pool, the corresponding fees are first deducted from the amount $a$, and then the remaining amount of token $X$ is traded as if there were no fees. Therefore, if $\phi \in [0, 1)$ is the fee that is charged, it is not difficult to check that, with the previous notations, the spot price formula of token $o$ in terms of token $i$ and taking the fee into consideration is

$$SP_{o,i}^{\text{fee}} = \frac{1}{1-\phi} \cdot \frac{\dfrac{B_i}{W_i}}{\dfrac{B_o}{W_o}}.$$

This is what we find in the Balancer Math code[1] and is shown in Listing 3-1.

**Listing 3-1.** Computation of the spot price in the Balancer code

```
/**********************************************************
// calcSpotPrice                                        //
// sP = spotPrice                                       //
// bI = tokenBalanceIn            (bI/wI)          1 //
// bO = tokenBalanceOut    sP = ------- * ------ //
// wI = tokenWeightIn             (bO/wO)      (1-sF) //
// wO = tokenWeightOut                                  //
// sF = swapFee                                         //
**********************************************************/
function calcSpotPrice(
    uint tokenBalanceIn,
    uint tokenWeightIn,
    uint tokenBalanceOut,
    uint tokenWeightOut,
    uint swapFee
)
```

---

[1] https://github.com/balancer-labs/balancer-core/blob/master/contracts/BMath.sol

```
    public pure
    returns (uint spotPrice)
{
    uint numer = bdiv(tokenBalanceIn, tokenWeightIn);
    uint denom = bdiv(tokenBalanceOut,
↪  tokenWeightOut);
    uint ratio = bdiv(numer, denom);
    uint scale = bdiv(BONE, bsub(BONE, swapFee));
    return (spotPrice = bmul(ratio, scale));
}
```

**Example 3.1.** Consider a Balancer pool with tokens ETH and USDC with the following balances and weights:

|       | ETH    | USDC       |
|-------|--------|------------|
| $B_j$ | 10,000 | 10,000,000 |
| $W_j$ | 0.8    | 0.2        |

Using Equation 3.1, we find that the spot price of ETH in terms of USDC is

$$SP_{ETH,USDC} = \frac{\dfrac{10,000,000}{0.2}}{\dfrac{10,000}{0.8}} = 4,000.$$

**Example 3.2.** Consider a Balancer pool of two tokens $X$ and $Y$, with weights and balances given by

|       | X   | Y   |
|-------|-----|-----|
| $B_j$ | 10  | 100 |
| $W_j$ | 0.5 | 0.5 |

Hence, token $X$ is ten times more valuable than token $Y$ since the spot price of token $X$ in terms of token $Y$ is 10.

If, instead of having the same weights, the weights had been given by the following table, then the spot price of token $X$ in terms of token $Y$ would have been 1.

|        | X              | Y               |
|--------|----------------|-----------------|
| $B_j$  | 10             | 100             |
| $W_j$  | $\dfrac{1}{11}$ | $\dfrac{10}{11}$ |

In other words, the tokens $X$ and $Y$ become equally valued since the weight of token $Y$ is now ten times the weight of token $X$.

As we can see, the weights give the necessary flexibility so that the reserves of the different tokens in the pool do not need to have the same value. Clearly, in the second case, the reserves of token $Y$ of the pool (100 tokens $Y$) are worth ten times more than the reserves of token $X$ of the pool (10 tokens $X$) since 1 token $X$ is worth 1 token $Y$. This can be seen in the weights, since the weight of token $Y$ in the pool is ten times the weight of token $X$.

## 3.2.1 Constant Value Distribution

We will prove now that in any Balancer pool, the share of a token in the pool is always its weight.

Consider a Balancer's liquidity pool with $N$ tokens with balances $B_1$, $B_2$, ..., $B_N$ and weights $W_1$, $W_2$, ..., $W_N$.

Let $j, k \in \{1, 2, ..., N\}$. Observe that the value of $B_j$ tokens $j$ in terms of token $k$ is $B_j SP_{j,k}$. Thus, the value of the whole pool in terms of token $k$ is

$$V_k = \sum_{j=1}^{N} B_j SP_{j,k} = \sum_{j=1}^{N} B_j \frac{B_k W_j}{W_k B_j} = \sum_{j=1}^{N} \frac{B_k W_j}{W_k} = \frac{B_k}{W_k} \sum_{j=1}^{N} W_j = \qquad (3.2)$$
$$= \frac{B_k}{W_k}.$$

And since the value of all the tokens $k$ of the pool in terms of token $k$ itself is $B_k$, we obtain that the share of token $k$ in the pool is

$$\frac{B_k}{\dfrac{B_k}{W_k}} = W_k$$

and it does not depend on the state of the pool. That is, the share of token $k$ in the pool is always the weight $W_k$.

It is interesting to observe that if we consider a Balancer pool with two equally weighted tokens, then, for $k \in \{1, 2\}$, the value of the pool in terms of token $k$ is $V_k = \dfrac{B_k}{0.5} = 2B_k$, and hence, we obtain the formula for the value of a Uniswap v2 pool given in subsection 2.3.3.

**Example 3.3.** Consider a Balancer pool with tokens ETH, MATIC, and USDC, whose balances are given by the following:

|       | ETH | MATIC     | USDC      |
|-------|-----|-----------|-----------|
| $B_j$ | 750 | 1,000,000 | 5,000,000 |

Suppose that the price of ETH in terms of USDC is 4,000 and that the price of MATIC in terms of USDC is 2. Then, the value of the whole pool is

$$750 \cdot 4,000 + 2 \cdot 1,000,000 + 5,000,000 = 10,000,000 \text{ USDC}$$

Hence, the weight of USDC in the pool is

$$\frac{5{,}000{,}000}{10{,}000{,}000} = 0.5.$$

On the other hand, the value of the 750 ETH of the pool in terms of USDC is 3,000,000, and thus, the weight of ETH in the pool is

$$\frac{3{,}000{,}000}{10{,}000{,}000} = 0.3.$$

In a similar way, the value of the 1,000,000 MATIC of the pool in terms of USDC is 2,000,000, and then the weight of MATIC in the pool is

$$\frac{2{,}000{,}000}{10{,}000{,}000} = 0.2.$$

Therefore, we obtain that the state of the pool is given by the following table:

|         | ETH  | MATIC     | USDC      |
|---------|------|-----------|-----------|
| $B_j$   | 750  | 1,000,000 | 5,000,000 |
| $W_j$   | 0.3  | 0.2       | 0.5       |

The reader can easily check that the spot prices of ETH and MATIC with respect to USDC are 4,000 and 2, respectively.

## 3.3 Trading Formulae

We will now derive the trading formulae for Balancer pools. As in the case of the Uniswap v2 AMM studied in the previous chapter, these fundamental formulae show which amount of a token $X$ a trader needs to deposit into the pool in order to receive a certain amount of another token $Y$ and which amount of token $Y$ they will receive if they deposit into the pool a certain amount of token $X$.

We assume first that the pool has no fee. As we did before, suppose that a trader wants to buy token $o$ in exchange of token $i$. For each $j \in \{1, 2, ..., N\}$, let $B_j$ be the balance of token $j$ before the trade and let $B_j^{'}$ be the balance of token $j$ after the trade. Let $A_o$ be the amount of token $o$ that the trader receives and let $A_i$ be the amount of token $i$ the trader deposits. Recall that $B_o^{'} = B_o - A_o$, $B_i^{'} = B_i + A_i$ and $B_j^{'} = B_j$ for all $j \in \{1, 2, ..., N\} - \{i, o\}$. Since the value function $V$ is constrained to a constant value $V_0$ during the trade, we have that

$$\prod_{j=1}^{N} B_j^{W_j} = V_0 = \prod_{j=1}^{N} \left( B_j^{'} \right)^{W_j}.$$

And since $B_j^{'} = B_j$ for all $j \in \{1, 2, ..., N\} - \{i, o\}$, we obtain that $B_o^{W_o} B_i^{W_i} = \left( B_o^{'} \right)^{W_o} \left( B_i^{'} \right)^{W_i}$. Hence,

$$B_o^{W_o} B_i^{W_i} = \left( B_o - A_o \right)^{W_o} \left( B_i + A_i \right)^{W_i}.$$

Thus,

$$\left( \frac{B_o - A_o}{B_o} \right)^{W_o} = \left( \frac{B_i}{B_i + A_i} \right)^{W_i}. \tag{3.3}$$

Hence,

$$1 - \frac{A_o}{B_o} = \left( \frac{B_i}{B_i + A_i} \right)^{\frac{W_i}{W_o}}.$$

and then the amount $A_o$ of token $o$ that the trader receives when selling an amount $A_i$ of token $i$ if there are no fees is

$$A_o = B_o \left( 1 - \left( \frac{B_i}{B_i + A_i} \right)^{\frac{W_i}{W_o}} \right).$$

In a similar way, isolating $A_i$ from Equation 3.3, we obtain that the amount $A_i$ of token $i$ that the trader has to deposit in order to receive an amount $A_o$ of token $o$ if there are no fees is

$$A_i = B_i \left( \left( \frac{B_o}{B_o - A_o} \right)^{\frac{W_o}{W_i}} - 1 \right).$$

Suppose now that the pool has a fee $\phi \in [0, 1)$. As mentioned previously, the fees are charged on the way in, which is equivalent to deducting the fees from the deposited amount and then performing the trade with the remaining amount as if there were no fees. Thus, the corresponding trading formulae can be obtained simply by replacing $A_i$ with $(1 - \phi)A_i$ in the previous formulae. Thus, the amount $A_o$ of token $o$ that the trader receives when depositing an amount $A_i$ of token $i$ is

$$A_o = B_o \left( 1 - \left( \frac{B_i}{B_i + (1 - \phi) A_i} \right)^{\frac{W_i}{W_o}} \right). \tag{3.4}$$

and the amount $A_i$ of token $i$ that the trader has to deposit in order to receive an amount $A_o$ of token $o$ is

$$A_i = \frac{B_i}{1 - \phi} \left( \left( \frac{B_o}{B_o - A_o} \right)^{\frac{W_o}{W_i}} - 1 \right). \tag{3.5}$$

These formulae are implemented in the Balancer Math code,[2] as we can see in Listings 3-2 and 3-3.

***Listing 3-2.*** Function `calcOutGivenIn` of the Balancer smart contract

```
/**********************************************
// calcOutGivenIn                            //
// aO = tokenAmountOut                       //
// bO = tokenBalanceOut                      //
// bI = tokenBalanceIn                       //
// aI = tokenAmountIn                        //
// wI = tokenWeightIn                        //
// wO = tokenWeightOut                       //
// sF = swapFee                              //
//                                           //
//            /     /      bI        \ (wI/wO)\ //
// aO = bO * / 1 - / -------------- /^      / //
//           \     \(bI+(aI*(1-sF))/       / //
//                                           //
**********************************************/

    function calcOutGivenIn(
        uint tokenBalanceIn,
        uint tokenWeightIn,
        uint tokenBalanceOut,
        uint tokenWeightOut,
        uint tokenAmountIn,
        uint swapFee
    )
        public pure
```

---

[2] https://github.com/balancer-labs/balancer-core/blob/master/contracts/BMath.sol

```
        returns (uint tokenAmountOut)
    {
        uint weightRatio = bdiv(tokenWeightIn,
↪   tokenWeightOut);
        uint adjustedIn = bsub(BONE, swapFee);
        adjustedIn = bmul(tokenAmountIn, adjustedIn);
        uint y = bdiv(tokenBalanceIn,
↪   badd(tokenBalanceIn, adjustedIn));
        uint foo = bpow(y, weightRatio);
        uint bar = bsub(BONE, foo);
        tokenAmountOut = bmul(tokenBalanceOut, bar);
        return tokenAmountOut;
    }
```

***Listing 3-3.*** Function `calcInGivenOut` of the Balancer smart contract

```
/*****************************************************
// calcInGivenOut                                    //
// aI = tokenAmount In                               //
// bO = tokenBalanceOut                              //
// bI =  tokenBalanceIn                              //
// aO = tokenAmountOut                               //
// wI =  tokenWeightIn                               //
// wO = tokenWeightOut                               //
// sF = swapFee                                      //
//                                                   //
//                 / /      bO      \  (wO/wI)    \   //
//         bI * / / ----------- /^        - 1 /       //
// aI =          \  \( bO - aO )/              /      //
//                                                   //
//         -------------------------------------     //
//                     ( 1 - sF )                    //
//                                                   //
*****************************************************/
```

```
function calcInGivenOut(
    uint tokenBalanceIn,
    uint tokenWeightIn,
    uint tokenBalanceOut,
    uint tokenWeightOut,
    uint tokenAmountOut,
    uint swapFee
)
    public pure
    returns (uint tokenAmountIn)
{
    uint weightRatio = bdiv(tokenWeightOut,
↪  tokenWeightIn);
    uint diff = bsub(tokenBalanceOut,
↪  tokenAmountOut);
    uint y = bdiv(tokenBalanceOut, diff);
    uint foo = bpow(y, weightRatio);
    foo = bsub(foo, BONE);
    tokenAmountIn = bsub(BONE, swapFee);
    tokenAmountIn = bdiv(bmul(tokenBalanceIn, foo),
↪  tokenAmountIn);
    return tokenAmountIn;
}
```

Observe that Equations 2.7 and 2.8 of the Uniswap v2 AMM are particular cases of Equations 3.4 and 3.5, corresponding to an equally weighted Balancer pool of only two tokens.

# 3.4  Providing Liquidity

As with Uniswap v2 pools, anyone can provide liquidity to a Balancer pool in order to earn a portion of the trading fees. In the case of the Balancer AMM, liquidity providers are given Balancer Pool Tokens (BPT) in return for their deposit. In the same way as the LP tokens of Uniswap v2 pools, the BPT are receipts representing ownership of a portion of the assets contained in a certain Balancer pool. Balancer allows two types of token asset deposit (minting BPT) and two types of token asset withdrawal (burning BPT):

- All-asset deposit
- Single-asset deposit
- All-asset withdrawal
- Single-asset withdrawal

We will cover all these cases in detail, starting, in the next subsection, with the all-asset deposit case.

# 3.4.1  All-Asset Deposit

In the all-asset deposit case, the liquidity provider deposits certain amounts of each of the tokens of the pool. These amounts have to be proportional to the corresponding balances of the tokens in the pool, and the liquidity provider will obtain BPT in that proportion.

In the following example, we will see that after an all-asset deposit, the value of Balancer's value function increases by the same proportion as the balances of the tokens of the pool.

**Example 3.4.** Consider a Balancer pool having three tokens, ETH, USDC, and DAI, with weights and balances given in the following table:

|         | Token 1 | Token 2   | Token 3   |
|---------|---------|-----------|-----------|
|         | ETH     | USDC      | DAI       |
| $B_j$   | 1,000   | 2,400,000 | 1,600,000 |
| $W_j$   | 0.5     | 0.3       | 0.2       |

The value of the Balancer's value function is

$$V_0 = B_1^{W_1} \cdot B_2^{W_2} \cdot B_3^{W_3} = 1,000^{0.5} \cdot 2,400,000^{0.3} \cdot 1,600,000^{0.2}$$
$$\approx 45,173.8774.$$

If a liquidity provider deposits 10% of the pool reserves—in this case, 100 ETH, 240,000 USDC, and 160,000 DAI—the updated value of Balancer's value function is

$$V_1 = 1,100^{0.5} \cdot 2,640,000^{0.3} \cdot 1,760,000^{0.2} \approx 49,691.2652$$

Note that

$$\frac{V_1}{V_0} = \frac{49,691.2652}{45,173.8774} = 1.1,$$

that is, the value of Balancer's value function increases by exactly 10%.

We will analyze now the general case. Suppose that we have a Balancer pool of $N$ tokens with balances $B_1, B_2, ..., B_N$ and weights $W_1, W_2, ..., W_N$. Let

$$V_0 = \prod_{j=1}^{N} B_j^{W_j}$$

and let $M$ be the total number of BPT in circulation. For each $j \in \{1, 2, ..., N\}$, let $a_j$ be the amount of token $j$ that the liquidity provider will deposit. Since the amounts $a_1, a_2, ..., a_N$ have to be proportional to the balances $B_1, B_2, ..., B_N$, we obtain that

$$\frac{a_1}{B_1} = \frac{a_2}{B_2} = \dots = \frac{a_N}{B_N}.$$

Let $q = \dfrac{a_1}{B_1}$; that is, $q$ is the proportion that will be added to the pool. Note that the liquidity provider will obtain $qM$ BPT.

We will now compute the change in the value of the value function $V$. Let $B'_1, B'_2, \dots, B'_N$ be the balances in the pool after the deposit. Note that for each $j \in \{1, 2, \dots, N\}$, we have that

$$\frac{B'_j}{B_j} = \frac{B_j + a_j}{B_j} = 1 + q$$

and hence $B'_j = (1+q)B_j$. Let $V_1$ be the value of the value function after the deposit. We have that

$$V_1 = \prod_{j=1}^{N}\left(B'_j\right)^{W_j} = \prod_{j=1}^{N}\left((1+q)B_j\right)^{W_j} = \prod_{j=1}^{N}(1+q)^{W_j} B_j^{W_j} =$$

$$= \prod_{j=1}^{N}(1+q)^{W_j} \prod_{j=1}^{N}B_j^{W_j} = (1+q)^{\sum_{j=1}^{N}W_j} V_0 = (1+q)V_0.$$

That is, the value of the value function increases in the same proportion $q$.

To sum up, we outline in the following table the amounts of tokens that need to be deposited to reach an increase of a proportion $q$ in the token balances and the amount of BPT received for such a deposit.

| | |
|---|---|
| Proportion | $q$ |
| Amount of token $j$ to deposit | $qB_j$ |
| Amount of BPT received | $qM$ |
| Updated value of V | $(1 + q)V_0$ |

We will prove now that the spot prices do not change after the deposit. Let $i, o \in \{1, 2, ..., N\}$ and let $SP_{o,i}$ and $SP'_{o,i}$ be the spot prices of token $o$ in terms of token $i$ before the deposit and after the deposit, respectively. We have that

$$SP'_{o,i} = \frac{\dfrac{B'_i}{W_i}}{\dfrac{B'_o}{W_o}} = \frac{\dfrac{(1+q)B_i}{W_i}}{\dfrac{(1+q)B_o}{W_o}} = \frac{\dfrac{B_i}{W_i}}{\dfrac{B_o}{W_o}} = SP_{o,i}.$$

That is, the spot prices before the deposit and after the deposit coincide.

## 3.4.2 Single-Asset Deposit

In the single-asset deposit case, the liquidity provider deposits a certain amount of only one token of the pool. Thus, we do not have a proportional deposit as in the previous case. Hence, the corresponding parameter $q$ and the amount of BPT to mint will have to be computed in a different way. In the previous case, we found out that if $V_0$ and $V_1$ are the values of the value function before and after the deposit, respectively, and $q$ is the proportion of the deposit, then $V_1 = (1 + q)V_0$. Thus, the parameter $q$ and the amount of BPT to be given to the liquidity provider can be obtained from the proportion in which the value of Balancer's value function has increased.

Another point has to be considered, though. Since the liquidity provider is depositing only one token, all the spot prices that are related to this token will change since in the quotient

$$SP_{o,i} = \frac{\dfrac{B_i}{W_i}}{\dfrac{B_o}{W_o}}$$

exactly one of the balances $B_i$, $B_o$ will change while the other will remain the same (see Example 3.5). Thus, it makes sense to charge a fee in order to compensate the current liquidity providers for possible impermanent losses following the price change. However, one can argue that if the new liquidity provider wants to deposit an amount $a_i$ of token $i$, then an amount $W_i a_i$ of token $i$ can be deposited as it is (since the weight $W_i$ coincides with the share of token $i$ in the pool), while the remaining amount $(1 - W_i)a_i$ should be converted to the other tokens and deposited in that way. Hence, Balancer decides to charge the fee $\phi$ only on the amount $(1 - W_i)a_i$, and thus, Balancer will consider that the new liquidity provider is depositing an amount $a_i - \phi(1 - W_i)a_i$ of token $i$ in order to compute how many BPT they have to be given.

Under these considerations, if the liquidity pool has $N$ tokens with weights $W_1, W_2, ..., W_N$ and balances $B_1, B_2, ..., B_N$ before the deposit and a new liquidity provider wants to deposit an amount $a_i$ of token $i$, the new computed value of Balancer's value function is

$$V_o' = \left( B_i + a_i - \phi\left(1 - W_i\right)a_i \right)^{W_i} \prod_{\substack{j=1 \\ j \neq i}}^{N} B_j^{W_j}$$

$$= \left( \frac{B_i + a_i - \phi\left(1 - W_i\right)a_i}{B_i} \right)^{W_i} \prod_{j=1}^{N} B_j^{W_j}$$

$$= \left( \frac{B_i + a_i - \phi\left(1 - W_i\right)a_i}{B_i} \right)^{W_i} V_o$$

Thus, we define

$$q = \left( \frac{B_i + a_i - \phi\left(1 - W_i\right)a_i}{B_i} \right)^{W_i} - 1 \qquad (3.6)$$

And if $M$ is the total number of BPT in existence before the deposit, the amount of BPT that is given to the new liquidity provider is

$$\left(\frac{B_i + a_i - \phi(1-W_i)a_i}{B_i}\right)^{W_i} M - M. \tag{3.7}$$

In addition, from Equation 3.6, we obtain that in order to deposit a share $q$ of the existent BPT, the liquidity provider has to deposit an amount

$$a_i = \frac{B_i(1+q)^{\frac{1}{W_i}} - B_i}{1 - \phi(1-W_i)} \tag{3.8}$$

of token $i$.

Observe also that the number $V_0'$ defined before is not the real value of Balancer's value function after the deposit. With the previous notations, after the deposit, the balance of token $i$ is $B_i + a_i$, and for each $j \in \{1, 2, ..., N\} - \{i\}$, the balance of token $j$ is $B_j$ (the same as before the deposit). Thus, the value of Balancer's value function after the deposit is

$$V_1 = (B_i + a_i)^{W_i} \prod_{\substack{j=1 \\ j \neq i}}^{N} B_j^{W_j}.$$

**Example 3.5.** Consider a Balancer liquidity pool whose tokens are ETH, MATIC, and USDC and whose weights and balances are given in the following table.

|  | Token 1 | Token 2 | Token 3 |
|---|---|---|---|
|  | ETH | MATIC | USDC |
| $B_j$ | 1,000 | 2,000,000 | 8,000,000 |
| $W_j$ | 0.25 | 0.25 | 0.5 |

Observe that the spot prices (without fees) of ETH and MATIC with respect to USDC are

$$SP_{\text{ETH,USDC}} = \frac{\dfrac{B_3}{W_3}}{\dfrac{B_1}{W_1}} = \frac{\dfrac{8{,}000{,}000}{0.5}}{\dfrac{1{,}000}{0.25}} = 4{,}000$$

and

$$SP_{\text{MATIC,USDC}} = \frac{\dfrac{B_3}{W_3}}{\dfrac{B_2}{W_2}} = \frac{\dfrac{8{,}000{,}000}{0.5}}{\dfrac{2{,}000{,}000}{0.25}} = 2$$

respectively, and the spot price of ETH with respect to MATIC is

$$SP_{\text{ETH,MATIC}} = \frac{\dfrac{B_2}{W_2}}{\dfrac{B_1}{W_1}} = \frac{\dfrac{2{,}000{,}000}{0.25}}{\dfrac{1{,}000}{0.25}} = 2{,}000.$$

Suppose now that a liquidity provider deposits 50 ETH into the pool. The spot prices (without fees) of ETH and MATIC with respect to USDC are now

$$SP_{\text{ETH,USDC}} = \frac{\dfrac{B_3}{W_3}}{\dfrac{B_1}{W_1}} = \frac{\dfrac{8{,}000{,}000}{0.5}}{\dfrac{1{,}050}{0.25}} \approx 3{,}809.52$$

and

$$SP_{\text{MATIC,USDC}} = \frac{\dfrac{B_3}{W_3}}{\dfrac{B_2}{W_2}} = \frac{\dfrac{8{,}000{,}000}{0.5}}{\dfrac{2{,}000{,}000}{0.25}} = 2$$

respectively, and the spot price of ETH with respect to MATIC is now

$$SP_{\text{ETH,MATIC}} = \frac{\dfrac{B_2}{W_2}}{\dfrac{B_1}{W_1}} = \frac{\dfrac{2{,}000{,}000}{0.25}}{\dfrac{1{,}050}{0.25}} \approx 1{,}904.76.$$

Observe that the spot price of ETH with respect to USDC and MATIC has changed, since the balance of ETH in the pool has changed, while the balances of MATIC and USDC have not. In addition, this last fact implies that the spot price of MATIC with respect to USDC did not change, as we can see.

**Example 3.6.** As in the previous example, consider a Balancer liquidity pool with tokens ETH, MATIC, and USDC having weights and balances given in the following table:

| | Token 1<br>ETH | Token 2<br>MATIC | Token 3<br>USDC |
|---|---|---|---|
| $B_j$ | 1,000 | 2,000,000 | 8,000,000 |
| $W_j$ | 0.25 | 0.25 | 0.5 |

As usual, we will assume that the liquidity pool has a fee $\phi = 0.003$. Suppose now that, as in the previous example, a liquidity provider deposits 50 ETH into the pool and that there were 1,000,000 BPT before the new deposit. Applying Equation 3.7, we obtain that the amount of BPT that the new liquidity provider receives is

$$\left(\frac{B_1 + a_1 - \phi(1 - W_1)a_1}{B_1}\right)^{W_1} M - M$$

$$= \left(\frac{1,000 + 50 - 0.003(1 - 0.25)50}{1,000}\right)^{0.25} \cdot 1,000,000 - 1,000,000$$

$$\approx 12,245.12.$$

Hence, after the deposit, there will be (approximately) a total of 1,012,245.12 BPT.

We will now analyze the situation of another liquidity provider who had 1,000 BPT before the deposit of 50 ETH that we considered before. In the following table, we show the value of these 1,000 BPT in terms of USDC and ETH before and after the new deposit.

|  | Before the new deposit | After the new deposit |
| --- | --- | --- |
| Amount of BPT | 1,000 | 1,000 |
| Total BPT | 1,000,000 | 1,012,245.12 |
| Share | 0.001 | 0.000987903 |
| Total pool value (in USDC) | 16,000,000 | 16,000,000 |
| Value of 1,000 BPT (in USDC) | 16,000 | 15,806.45 |
| Total pool value (in ETH) | 4,000 | 4,200 |
| Value of 1,000 BPT (in ETH) | 4 | 4.15 |

Recall that for computing the total pool values, we are applying Equation 3.2. As we can see, the value in USDC of the 1,000 BPT held by the liquidity provider has decreased after the new deposit, but the value in ETH of their position has increased. Observe that the spot price of ETH in terms of USDC before the new deposit was 4,000 and after this deposit is 3,809.52, as we have seen in Example 3.5.

We will now compare the single-asset deposit with the all-asset deposit. The following proposition states that in the zero-fee case, a single-asset deposit of an amount $a$ is equivalent to trading portions of the amount $a$ to obtain adequate amounts of all the tokens of the pool and then performing an all-asset deposit with those amounts.

**Proposition 3.1.** *Consider a Balancer liquidity pool with N tokens and no fees. Let $q > 0$ and let $k \in \{1, 2, ..., N\}$. Suppose that a new liquidity provider wants to deposit a share q of the pool using only token k. We consider the following two ways of achieving that:*

   (1)   *Performing a single-asset deposit of a suitable amount of token k*

   (2)   *Performing N – 1 trades with token k to obtain adequate amounts of all the tokens different from k and then doing an all-asset deposit*

*Then, the amount of token k needed for process 1 is the same as the amount needed for process 2.*

*Proof.* Let $B_1, B_2, ..., B_N$ be the balances of tokens 1, 2, ..., $N$ before any of these processes and let $W_1, W_2, ..., W_N$ be the corresponding weights. Let $V_0 = \prod_{j=1}^{N} B_j^{W_j}$. For $l \in \{1, 2\}$, let $A_l$ be the amount of token $k$ needed for process $l$ and let $V_l$ be the value of Balancer's value function after process $l$ has been completed.

We will analyze process 1 first. From Equation 3.6, we obtain that

$$q = \left( \frac{B_k + A_1}{B_k} \right)^{W_k} - 1.$$

Hence,

$$V_1 = \left( B_k + A_1 \right)^{W_k} \prod_{\substack{j=1 \\ j \neq k}}^{N} B_j^{W_j} = \left( 1 + q \right) B_k^{W_k} \prod_{\substack{j=1 \\ j \neq k}}^{N} B_j^{W_j} = \left( 1 + q \right) V_0.$$

Now, we will analyze process 2. Given positive real numbers $a_1$, $a_2, ..., a_N$, we will perform the $N - 1$ trades described in the following text. For each $j \in \{1, 2, ..., N\} - \{k\}$, we define trade $j$ as follows: we deposit an amount $a_j$ of token $k$ and obtain an amount $a'_j$ of token $j$. These trades can be performed in any order, but for simplicity, suppose that they are performed in increasing order following the index $j$. After these $N - 1$ trades have been done, the balance of token $k$ in the pool is $B_k + \displaystyle\sum_{\substack{j=1 \\ j \neq k}}^{N} a_j$,

and for $j \in \{1, 2, ..., N\} - \{k\}$, the balance of token $j$ in the pool is $B_j - a'_j$. Observe that the value of the value function after these $N - 1$ trades is still $V_0$ since we are assuming that the pool has no fees.

Now we want to do an all-asset deposit of amounts $a_k$ for token $k$ and $a'_j$ for all tokens $j$ with $j \neq k$. Moreover, this deposit should represent a share $q$. To this end, we need that

$$\frac{a'_j}{B_j - a'_j} = q$$

for all $j \in \{1, 2, ..., N\} - \{k\}$, and

$$\frac{a_k}{B_k + \displaystyle\sum_{\substack{j=1 \\ j \neq k}}^{N} a_j} = q.$$

Observe that the numbers $a'_j$, for $j \neq k$, can be computed from the previous equations by

$$a'_j = \frac{qB_j}{1+q},$$

then the amounts $a_j$, for $j \neq k$, can be obtained using Equation 3.5, and finally, the amount $a_k$ can be computed. Note also that $A_2 = \sum_{j=1}^{N} a_j$.

Note that after the all-asset deposit described previously, the balance of token $k$ in the pool is $B_k + \sum_{j=1}^{N} a_j = B_k + A_2$, and for all $j \in \{1, 2, ..., N\} - \{k\}$, the balance of token $j$ in the pool is $B_j$. Hence,

$$V_2 = \left(B_k + A_2\right)^{W_k} \prod_{\substack{j=1 \\ j \neq k}}^{N} B_j^{W_j}.$$

Recall also that from Subsection 3.4.1, we know that $V_2 = (1 + q)V_0$. Thus,

$$V_1 = \left(B_k + A_1\right)^{W_k} \prod_{\substack{j=1 \\ j \neq k}}^{N} B_j^{W_j} = (1 + q)V_0 = V_2 = \left(B_k + A_2\right)^{W_k} \prod_{\substack{j=1 \\ j \neq k}}^{N} B_j^{W_j}$$

and hence, $\left(B_k + A_1\right)^{W_k} = \left(B_k + A_2\right)^{W_k}$. Therefore, $A_1 = A_2$. $\square$

The following is a numerical example of the result of the previous proposition.

**Example 3.7.** Consider a Balancer liquidity pool with no fees and three tokens: ETH, MATIC, and USDC. Suppose that the respective weights and balances are given in the following table:

|  | Token 1 ETH | Token 2 MATIC | Token 3 USDC |
|---|---|---|---|
| $B_j$ | 1,100 | 2,200,000 | 8,800,000 |
| $W_j$ | 0.25 | 0.25 | 0.5 |

Suppose that a new liquidity provider wants to deposit a share $q = 0.1$ of the pool using only ETH. The first way of doing that will be performing a single-asset deposit of an amount $A_1$ of ETH satisfying that

$$q = \left( \frac{B_1 + A_1}{B_1} \right)^{W_k} - 1,$$

that is,

$$0.1 = \left( \frac{1,100 + A_1}{1,100} \right)^{0.25} - 1.$$

Thus, we obtain that $A_1 = 510.51$.

Another way to do that will be to perform one ETH/MATIC trade and one ETH/USDC trade in order to obtain the needed amounts of MATIC and USDC, and then perform an all-asset deposit. As in the proof of the previous proposition, we will compute the amounts of MATIC and USDC that are needed, and then we will find out how much ETH we need to get those. With the notations of the previous proof, we have that

$$\frac{a_2'}{B_2 - a_2'} = 0.1 \text{ and } \frac{a_3'}{B_3 - a_3'} = 0.1 .$$

Thus, $a_2' = 200,000$ and $a_3' = 800,000$. In order to obtain 200,000 MATIC, the liquidity provider needs to trade the following amount of ETH:

$$a_2 = 1,100 \left( \left( \frac{2,200,000}{2,200,000 - 200,000} \right)^{\frac{0.25}{0.25}} - 1 \right) = 110.$$

And the balances of the pool after the trade are as follows:

| ETH | MATIC | USDC |
|---|---|---|
| 1,210 | 2,000,000 | 8,800,000 |

Then, in order to obtain 800,000 USDC, the liquidity provider needs to trade the following amount of ETH:

$$a_3 = 1,210 \left( \left( \frac{8,800,000}{8,800,000 - 800,000} \right)^{\frac{0.5}{0.25}} - 1 \right) = 254.1.$$

And the balances of the pool after the trade are as follows:

| ETH | MATIC | USDC |
|---|---|---|
| 1,464.1 | 2,000,000 | 8,000,000 |

Now, to make the all-asset deposit, the liquidity provider needs an amount $a_1$ of ETH such that $\frac{a_1}{1,464.1} = q = 0.1$. Hence, they need 146.41 ETH. Thus, they can perform an all-asset deposit of 146.41 ETH, 200,000 MATIC, and 800,000 USDC. The balances of the pool after the all-asset deposit are as follows:

| ETH | MATIC | USDC |
|---|---|---|
| 1,610.51 | 2,200,000 | 8,800,000 |

which are the same as the balances after the single-asset deposit considered previously. Note also that the amount of ETH needed in this second process is $A_2 = 110 + 254.1 + 146.41 = 510.51$, and hence, $A_2 = A_1$.

In the following example, we will show that the previous proposition does not hold if the pool has nonzero fees.

**Example 3.8.** Consider a Balancer liquidity pool with tokens ETH, MATIC, and USDC and with fee $\phi = 0.003$. Suppose that the weights and balances of the tokens are given in the following table:

|       | ETH   | MATIC     | USDC      |
|-------|-------|-----------|-----------|
| $B_j$ | 875   | 1,750,000 | 7,000,000 |
| $W_j$ | 0.25  | 0.25      | 0.5       |

Suppose that a new liquidity provider wants to deposit a share $q = 0.4$ of the pool using only ETH. From Equation 3.8, we obtain that in order to achieve that, the liquidity provider needs to deposit an amount of

$$A_1 = \frac{875(1+0.4)^{\frac{1}{0.25}} - 875}{1-0.003(1-0.25)} \approx 2{,}492.007 \quad \text{ETH.}$$

If the liquidity provider wants instead to make an all-asset deposit into the pool, they need to make one ETH/MATIC trade and one ETH/USDC trade so as to obtain the required amounts $a_2'$ and $a_3'$ of MATIC and USDC, respectively, which must satisfy $\dfrac{a_2'}{B_2 - a_2'} = 0.4$ and $\dfrac{a_3'}{B_3 - a_3'} = 0.4$.

Thus, they need 500,000 MATIC and 2,000,000 USDC. To obtain the required amount of MATIC, they need to trade

$$a_2 = \frac{875}{1-0.003}\left(\left(\frac{1,750,000}{1,750,000-500,000}\right)^{\frac{0.25}{0.25}} - 1\right)$$

$$\approx 351.05316 \text{ETH,}$$

and the balances of the pool after the trade are as follows:

| ETH          | MATIC     | USDC      |
|--------------|-----------|-----------|
| 1,226.05316  | 1,250,000 | 7,000,000 |

Then, to obtain the required amount of USDC, the liquidity provider needs to trade

$$a_3 = \frac{1{,}226.05316}{1-0.003}\left(\left(\frac{7{,}000{,}000}{7{,}000{,}000-2{,}000{,}000}\right)^{\frac{0.5}{0.25}}-1\right)$$

$$\approx 1{,}180.55269 \text{ ETH},$$

and the balances of the pool after the trade are as follows:

| ETH | MATIC | USDC |
|---|---|---|
| 2,406.60585 | 1,250,000 | 5,000,000 |

Now, to make the all-asset deposit, the liquidity provider needs an amount $a_1$ of ETH such that $\dfrac{a_1}{2{,}406.60585} = q = 0.4$. Hence, they need 962.64234 ETH. Thus, they can perform an all-asset deposit of 962.64234 ETH, 500,000 MATIC, and 2,000,000 USDC. We can observe that the amount of ETH needed in this second process is

$$A_2 = 351.05316 + 1{,}180.55269 + 962.64234 = 2{,}494.24819,$$

which is greater than the amount $A_1 = 2{,}492.007$ needed for the single-asset deposit.

We will prove now that what we saw in the previous example is valid in general: when a Balancer pool has a nonzero fee, the amount of a certain token $X$ that is needed for a single-asset deposit is smaller than the amount of token $X$ needed to make the necessary trades and then perform an all-asset deposit. To this end, we need to give a couple of lemmas.

**Lemma 3.1.** *Consider a Balancer's liquidity pool with N tokens and fee $\phi$. For each $j \in \{1, 2, ..., N\}$, let $B_j$ be the balance of token $j$ and let $W_j$ be its weight. Let $i, o \in \{1, 2, ..., N\}$. Suppose that a trader deposits an amount $A_i$ of token $i$ into the pool and receives an amount $A_o$ of token $o$. Let $V_0$ and $V_1$ be the values of Balancer's value function before and after the trade, respectively. Then*

$$V_1 = V_0 \left( \frac{B_o - A_o}{B_o} \right)^{W_o} \left( \frac{\left( \frac{B_o}{B_o - A_o} \right)^{\frac{W_o}{W_i}} - \phi}{1 - \phi} \right)^{W_i}$$

*Proof.* Note that the balance of token $o$ after the trade is $B_o - A_o$, the balance of token $i$ after the trade is $B_i + A_i$, and for all $j \in \{1, 2, ..., N\} - \{i, o\}$, the balance of token $j$ after the trade is $B_j$. Thus,

$$V_1 = (B_o - A_o)^{W_o} (B_i + A_i)^{W_i} \prod_{\substack{j=1 \\ j \neq i,o}}^{N} B_j^{W_j}$$

$$= (B_o - A_o)^{W_o} (B_i + A_i)^{W_i} \frac{V_0}{B_o^{W_o} B_i^{W_i}}$$

$$= V_0 \left( \frac{B_o - A_o}{B_o} \right)^{W_o} \left( \frac{B_i + A_i}{B_i} \right)^{W_i}.$$

On the other hand, from Equation 3.5, we know that

$$A_i = \frac{B_i}{1 - \phi} \left( \left( \frac{B_o}{B_o - A_o} \right)^{\frac{W_o}{W_i}} - 1 \right)$$

Hence,

$$\frac{B_i + A_i}{B_i} = \frac{B_i + \frac{B_i}{1 - \phi} \left( \left( \frac{B_o}{B_o - A_o} \right)^{\frac{W_o}{W_i}} - 1 \right)}{B_i}$$

$$= 1 + \frac{1}{1-\phi}\left(\left(\frac{B_o}{B_o - A_o}\right)^{\frac{W_o}{W_i}} - 1\right)$$

$$= \frac{1}{1-\phi}\left(1 - \phi + \left(\frac{B_o}{B_o - A_o}\right)^{\frac{W_o}{W_i}} - 1\right)$$

$$= \frac{1}{1-\phi}\left(\left(\frac{B_o}{B_o - A_o}\right)^{\frac{W_o}{W_i}} - \phi\right).$$

The result follows.                                                $\square$

The following technical lemma will also be needed in the proof of Proposition 3.2.

**Lemma 3.2.** *Let $x \in \mathbb{R}$ such that $x > 1$. Let $\alpha$, $\beta$, $\phi \in \mathbb{R}$ such that $\alpha > 0$, $\beta > 0$, and $0 \le \phi < 1$. Then*

$$\frac{x^\alpha - \phi}{1-\phi} \cdot \frac{x^\beta - \phi}{1-\phi} \ge \frac{x^{\alpha+\beta} - \phi}{1-\phi}$$

*and the equality holds if and only if $\phi = 0$.*

*Proof.* For all $x$, $\alpha$, $\beta$, $\phi$ satisfying the hypotheses of the lemma, let

$$L(x,\alpha,\beta,\phi) = \frac{x^\alpha - \phi}{1-\phi} \cdot \frac{x^\beta - \phi}{1-\phi} - \frac{x^{\alpha+\beta} - \phi}{1-\phi}.$$

Observe that

$$L(x,\alpha,\beta,\phi) = \frac{1}{(1-\phi)^2}\left(\left(x^\alpha - \phi\right)\left(x^\beta - \phi\right) - \left(x^{\alpha+\beta} - \phi\right)(1-\phi)\right)$$

$$= \frac{1}{\left(1-\phi\right)^2}\left(\phi x^{\alpha+\beta} - \phi x^{\alpha} - \phi x^{\beta} + \phi\right)$$

$$= \frac{\phi}{\left(1-\phi\right)^2}\left(x^{\alpha}-1\right)\left(x^{\beta}-1\right).$$

Since $x > 1$ and $\alpha, \beta > 0$, we obtain that $(x^{\alpha} - 1)(x^{\beta} - 1) > 0$, and since $\phi < 1$, we get $(1 - \phi)^2 > 0$. Thus, $L(x, \alpha, \beta, \phi) \geq 0$ since $\phi \geq 0$. In addition, $L(x, \alpha, \beta, \phi) = 0$ if and only if $\phi = 0$. The result follows.    □

Applying an inductive argument, we can extend the previous lemma in the following way.

**Corollary 3.1.** *Let $x \in \mathbb{R}$ such that $x > 1$, and let $n \in \mathbb{N}$ such that $n \geq 2$. Let $\alpha_1, \alpha_2, ..., \alpha_n \in \mathbb{R}$ such that $\alpha_j > 0$ for all $j \in \{1, 2, ..., n\}$. Let $\phi \in \mathbb{R}$ such that $0 \leq \phi < 1$. Then*

$$\prod_{j=1}^{n} \frac{x^{\alpha_j} - \phi}{1-\phi} \geq \frac{x^{\sum_{j=1}^{n}\alpha_j} - \phi}{1-\phi}$$

*and the equality holds if and only if $\phi = 0$.*

*Proof.* We proceed by induction on $n$. If $n = 2$, then the result holds by the previous lemma. Now, let $k \in \mathbb{N}$ such that $k \geq 2$ and suppose the the result holds for $n = k$. We will prove that the result holds for $n = k + 1$. Let $\alpha_1, \alpha_2, ..., \alpha_{k+1} \in \mathbb{R}$ such that $\alpha_j > 0$ for all $j \in \{1, 2, ..., k+1\}$. We have that

$$\prod_{j=1}^{k+1} \frac{x^{\alpha_j} - \phi}{1-\phi} = \left(\prod_{j=1}^{k} \frac{x^{\alpha_j} - \phi}{1-\phi}\right)\frac{x^{\alpha_{k+1}} - \phi}{1-\phi}$$

$$\geq \left(\frac{x^{\sum_{j=1}^{k}\alpha_j} - \phi}{1-\phi}\right)\frac{x^{\alpha_{k+1}} - \phi}{1-\phi} \geq \frac{x^{\sum_{j=1}^{k+1}\alpha_j} - \phi}{1-\phi},$$

where the first inequality holds by the inductive hypothesis (note that $1 - \phi > 0$ and $x^{\alpha_{k+1}} - \phi > 0$ since $x^{\alpha_{k+1}} > 1 > \phi$) and the second inequality holds by the previous lemma. Moreover, by the inductive hypothesis and the previous lemma, in both of the previous inequalities, the equality holds if and only if $\phi = 0$. The result follows. $\square$

**Proposition 3.2.** *Consider a Balancer's liquidity pool with N tokens and fee* $\phi \in (0, 1)$. *Let* $q > 0$ *and let* $k \in \{1, 2, ..., N\}$. *Suppose that a new liquidity provider wants to deposit a share q of the pool using only token k. We consider the following two ways of achieving that:*

(1) *Performing a single-asset deposit of a suitable amount of token k*

(2) *Performing N - 1 trades with token k to obtain adequate amounts of all the tokens different from k and then doing an all-asset deposit*

*Then, the amount of token k needed for process 2 is greater than the amount needed for process 1.*

*Proof.* Let $B_1$, $B_2$, ..., $B_N$ be the balances of tokens 1, 2, ..., N before any of these processes and let $W_1$, $W_2$, ..., $W_N$ be the corresponding weights. Let $V_0 = \prod_{j=1}^{N} B_j^{W_j}$. For $l \in \{1, 2\}$, let $A_l$ be the amount of token $k$ needed for process $l$ and let $V_l$ be the value of Balancer's value function after process $l$ has been completed.

We will analyze process 1 first. From Equation 3.8, we obtain that

$$A_1 = \frac{B_k (1+q)^{\frac{1}{W_k}} - B_k}{1 - \phi(1-W_k)} = B_k \cdot \frac{(1+q)^{\frac{1}{W_k}} - 1}{1 - \phi(1-W_k)}.$$

Now we will analyze process 2. Given positive real numbers $a_1$, $a_2$, ..., $a_N$, we will perform the $N - 1$ trades described in the following text. For each $j \in \{1, 2, ..., N\} - \{k\}$, we define trade $j$ as follows: we deposit an amount $a_j$ of token $k$ and obtain an amount $a'_j$ of token $j$. Let $V_{0,j}$ and $V_{1,j}$ be the values of Balancer's value function before and after trade $j$, respectively. By Lemma 3.1, we have that

$$V_{1,j} = V_{0,j} \left( \frac{B_j - a'_j}{B_j} \right)^{W_j} \left( \frac{\left( \dfrac{B_j}{B_j - a'_j} \right)^{\frac{W_j}{W_k}} - \phi}{1 - \phi} \right)^{W_k}$$

Note that the factor by which the value changes does not depend on the balance of token $k$ (which changes after each of the $N - 1$ trades). Hence, the value of Balancer's value function after the $N - 1$ trades are performed does not depend on the order in which these trades are executed, and this value is

$$V_2 = V_0 \prod_{\substack{j=1 \\ j \neq k}}^{N} \left( \frac{B_j - a'_j}{B_j} \right)^{W_j} \prod_{\substack{j=1 \\ j \neq k}}^{N} \left( \frac{\left( \dfrac{B_j}{B_j - a'_j} \right)^{\frac{W_j}{W_k}} - \phi}{1 - \phi} \right)^{W_k}$$

$$= B_k^{W_k} \prod_{\substack{j=1 \\ j \neq k}}^{N} \left( B_j - a'_j \right)^{W_j} \prod_{\substack{j=1 \\ j \neq k}}^{N} \left( \frac{\left( \dfrac{B_j}{B_j - a'_j} \right)^{\frac{W_j}{W_k}} - \phi}{1 - \phi} \right)^{W_k} .$$

Note that after these $N-1$ trades have been performed, for each $j \in \{1, 2, \ldots, N\} - \{k\}$, the balance of token $j$ in the pool is $B_j - a'_j$. Let $B'_k$ be the balance of token $k$ after the $N-1$ trades. Clearly,

$$V_2 = \left(B'_k\right)^{W_k} \prod_{\substack{j=1 \\ j \neq k}}^{N} \left(B_j - a'_j\right)^{W_j}$$

Hence,

$$\left(B'_k\right)^{W_k} = B_k^{W_k} \prod_{\substack{j=1 \\ j \neq k}}^{N} \left( \frac{\left(\dfrac{B_j}{B_j - a'_j}\right)^{\frac{W_j}{W_k}} - \phi}{1 - \phi} \right)^{W_k}$$

and thus,

$$B'_k = B_k \prod_{\substack{j=1 \\ j \neq k}}^{N} \left( \frac{\left(\dfrac{B_j}{B_j - a'_j}\right)^{\frac{W_j}{W_k}} - \phi}{1 - \phi} \right).$$

In particular, the balance of token $k$ after the $N-1$ trades does not depend on the order in which the trades are performed, and the amount of token $k$ needed for all these trades, which is $B'_k - B_k$, is also independent of the order of the trades.

Now we want to do an all-asset deposit of amounts $a_k$ for token $k$ and $a'_j$ for all tokens $j$ with $j \neq k$. Moreover, this deposit should represent a share $q$. To this end, we need that

$$\frac{a'_j}{B_j - a'_j} = q$$

for $j \in \{1, 2, ..., N\} - \{k\}$, and

$$\frac{a_k}{B'_k} = q.$$

Thus, we obtain that

$$a'_j = \frac{qB_j}{1+q}$$

for $j \neq k$, and $a_k = qB'_k$. Hence, the amount of token $k$ needed for process 2 is

$$A_2 = \left(B'_k - B_k\right) + qB'_k = \left(q+1\right)B'_k - B_k.$$

On the other hand, since $\dfrac{a'_j}{B_j - a'_j} = q$ for $j \neq k$, we obtain that

$$1 + q = 1 + \frac{a'_j}{B_j - a'_j} = \frac{B_j}{B_j - a'_j}$$

for all $j \in \{1, 2, ..., N\} - \{k\}$. Thus,

$$A_2 = (q+1)B'_k - B_k = (q+1)B_k \prod_{\substack{j=1 \\ j \neq k}}^{N} \left( \frac{\left( \dfrac{B_j}{B_j - a'_j} \right)^{\frac{W_j}{W_k}} - \phi}{1-\phi} \right) - B_k$$

$$= B_k \left( (q+1)\prod_{\substack{j=1 \\ j \neq k}}^{N} \left( \frac{(1+q)^{\frac{W_j}{W_k}} - \phi}{1-\phi} \right) - 1 \right).$$

Therefore,

$$\frac{A_2}{A_1} = \left( (q+1)\prod_{\substack{j=1 \\ j \neq k}}^{N} \left( \frac{(1+q)^{\frac{W_j}{W_k}} - \phi}{1-\phi} \right) - 1 \right) \frac{1-\phi(1-W_k)}{(1+q)^{\frac{1}{W_k}} - 1}.$$

Observe that $\dfrac{A_2}{A_1}$ does not depend on the balances of the pool.
Note that $1 - \phi(1 - W_k) > 0$ (because $0 \leq \phi < 1$ and $0 < 1 - W_k < 1$) and
$(1+q)^{\frac{1}{W_k}} - 1 > 0$. Hence, applying Corollary 3.1, we obtain that

$$\frac{A_2}{A_1} \geq \left( (q+1)\left( \frac{(1+q)^{\sum_{\substack{j=1 \\ j \neq k}}^{N} \frac{W_j}{W_k}} - \phi}{1-\phi} \right) - 1 \right) \frac{1-\phi(1-W_k)}{(1+q)^{\frac{1}{W_k}} - 1}$$

$$= \left( (q+1)\left( \frac{(1+q)^{\frac{1-W_k}{W_k}} - \phi}{1-\phi} \right) - 1 \right) \frac{1-\phi(1-W_k)}{(1+q)^{\frac{1}{W_k}} - 1}$$

$$= \left( \frac{(q+1)\left((1+q)^{\frac{1}{W_k}-1}-\phi\right)}{1-\phi} - 1 \right) \frac{1-\phi(1-W_k)}{(1+q)^{\frac{1}{W_k}}-1}$$

$$= \frac{(1+q)^{\frac{1}{W_k}}-\phi(1+q)-1+\phi}{1-\phi} \cdot \frac{1-\phi(1-W_k)}{(1+q)^{\frac{1}{W_k}}-1}$$

$$= \frac{(1+q)^{\frac{1}{W_k}}-\phi q-1}{(1+q)^{\frac{1}{W_k}}-1} \cdot \frac{1-\phi(1-W_k)}{1-\phi}$$

$$= \left(1 - \frac{\phi q}{(1+q)^{\frac{1}{W_k}}-1}\right)\left(1+\frac{\phi W_k}{1-\phi}\right)$$

$$= \left(1 - \phi \cdot \frac{q}{(1+q)^{\frac{1}{W_k}}-1}\right)\left(1+W_k \cdot \frac{\phi}{1-\phi}\right).$$

Now let $f: [0, +\infty) \to \mathbb{R}$ be defined by $f(x)=(1+x)^{\frac{1}{W_k}}$. Since $f$ is a continuous function that is differentiable in $(0, +\infty)$ and since $q > 0$, applying the Mean Value Theorem, we obtain that there exists $c \in (0, q)$ such that

$$\frac{f(q)-f(0)}{q-0}=f'(c),$$

that is,

$$\frac{(1+q)^{\frac{1}{W_k}}-1}{q}=\frac{1}{W_k}(1+c)^{\frac{1}{W_k}-1}.$$

Since $0 < W_k < 1$, we get that $\dfrac{1}{W_k} - 1 > 0$, and hence, $(1+c)^{\frac{1}{W_k}-1} > 1$ since $c > 0$. Thus,

$$\frac{(1+q)^{\frac{1}{W_k}} - 1}{q} > \frac{1}{W_k},$$

and hence,

$$\frac{q}{(1+q)^{\frac{1}{W_k}} - 1} < W_k.$$

Therefore,

$$\frac{A_2}{A_1} \geq \left(1 - \phi \cdot \frac{q}{(1+q)^{\frac{1}{W_k}} - 1}\right)\left(1 + W_k \cdot \frac{\phi}{1-\phi}\right)$$

$$> \left(1 - \phi W_k\right)\left(1 + W_k \cdot \frac{\phi}{1-\phi}\right)$$

$$= 1 + W_k \cdot \frac{\phi}{1-\phi} - W_k \phi - W_k^2 \cdot \frac{\phi^2}{1-\phi}$$

$$= 1 + W_k \phi \left(\frac{1}{1-\phi} - 1\right) - W_k^2 \cdot \frac{\phi^2}{1-\phi}$$

$$= 1 + W_k \phi \frac{\phi}{1-\phi} - W_k^2 \cdot \frac{\phi^2}{1-\phi}$$

$$= 1 + W_k \frac{\phi^2}{1-\phi}(1 - W_k) > 1$$

since $0 < W_k < 1$ and $0 < \phi < 1$. The result follows.  $\square$

# 3.4.3 All-Asset Withdrawal

We will now analyze the case in which a liquidity provider wants to redeem their BPT and receive the corresponding share of each of the tokens in the pool. Suppose that the liquidity pool has $N$ tokens with balances $B_1$, $B_2$, ..., $B_N$ and weights $W_1$, $W_2$, ..., $W_N$. Let $M$ be the total amount of BPT in existence and let $m$ be the amount of BPT the liquidity provider wants to exchange. This amount $m$ represents a fraction $\dfrac{m}{M}$ of the whole amount of BPT in existence. Thus, in exchange of the $m$ BPT (which will be burned), the liquidity provider will receive for each $j \in \{1, 2, ..., N\}$ an amount $\dfrac{m}{M} B_j$ of token $j$.

We will now study how the value of Balancer's value function changes when the $m$ BPT tokens are redeemed. Let

$$V_0 = \prod_{j=1}^{N} B_j^{W_j}$$

Note that $V_0$ is the value of Balancer's value function before the liquidity provider redeemed their BPT. Let $B'_1, B'_2, ..., B'_N$ be the balances in the pool after the exchange of the BPT. Note that for each $j \in \{1, 2, ..., N\}$, we have that

$$\frac{B'_j}{B_j} = \frac{B_j - \dfrac{m}{M} B_j}{B_j} = 1 - \frac{m}{M}.$$

Let $r = 1 - \dfrac{m}{M}$. Hence, $B'_j = rB_j$. Let $V_1$ be the value of Balancer's value function after the exchange of the BPT. We have that

$$V_1 = \prod_{j=1}^{N}\left(B'_j\right)^{W_j} = \prod_{j=1}^{N}\left(rB_j\right)^{W_j} = \prod_{j=1}^{N} r^{W_j} B_j^{W_j} = \prod_{j=1}^{N} r^{W_j} \prod_{j=1}^{N} B_j^{W_j}$$

$$= r^{\sum_{j=1}^{N} W_j} V_0 = rV_0 = \left(1 - \frac{m}{M}\right)V_0.$$

That is, the value of Balancer's value function decreases in the proportion $\frac{m}{M}$ after the liquidity provider redeems their BPT.

We outline in the following table the amounts of tokens that the liquidity provider will receive when redeeming their BPT and how the value of Balancer's value function is updated.

| | |
|---|---|
| Total amount of BPT in circulation | $M$ |
| Amount of BPT to redeem | $m$ |
| Amount of token $j$ to receive | $\dfrac{m}{M}B_j$ |
| Amount of BPT to be burned | $M$ |
| Updated value of $V$ | $\left(1 - \dfrac{m}{M}\right)V_0$ |
| Updated amount of BPT in circulation | $M - m$ |

## 3.4.4 Single-Asset Withdrawal

We will now analyze the case in which a liquidity provider wants to obtain only one pool token in return for their BPT. We will proceed in a similar way as we did for the single-asset deposit case.

Assume that the liquidity pool has $N$ tokens whose weights are $W_1$, $W_2$, ..., $W_N$. Let $B_1$, $B_2$, ..., $B_N$ be the balances before the withdrawal and let $\phi$ be the pool fee. Suppose that a liquidity provider has a share $\sigma$ of BPT and wants to withdraw only token $i$. Let $a_i$ be the amount of token $i$ that the liquidity provider would receive if there were no fees. On the other hand, in the previous case, we found out that if $V_0$ and $V_1$ are the values of Balancer's value function before and after the withdrawal, respectively, then $V_1 = (1 - \sigma)V_0$. Using this relation for this case, we obtain that

$$(1-\sigma)V_0 = V_1 = (B_i - a_i)^{W_i} \prod_{\substack{j=1 \\ j \neq i}}^{N} B_j^{W_j} = \left( \frac{B_i - a_i}{B_i} \right)^{W_i} V_0$$

and hence,

$$1 - \sigma = \left( \frac{B_i - a_i}{B_i} \right)^{W_i}.$$

Then,

$$a_i = B_i - B_i (1-\sigma)^{\frac{1}{W_i}}. \tag{3.9}$$

In the case that there is a nonzero fee, Balancer will charge the fee on the amount of token $i$ that the liquidity provider receives, and thus, they will receive an amount $a_i'$, which is smaller than $a_i$. As in the single-asset deposit, we deem that the fee should be charged on the amount they receive that would not correspond to token $i$ if we were to consider an all-asset withdrawal. Hence, the fee is charged on the amount $(1 - W_i)a_i$, and thus,

$$a_i' = a_i - \phi(1-W_i)a_i = (1-\phi(1-W_i))a_i.$$

Therefore, applying Equation 3.9, we obtain that

$$a_i' = (1-\phi(1-W_i))\left( B_i - B_i (1-\sigma)^{\frac{1}{W_i}} \right).$$

Observe that the number $V_1$ defined before is not the real value of Balancer's value function after the withdrawal. With the previous notations, the balance of token $i$ after the withdrawal is $B_i - a'_i$, and for each $j \in \{1, 2, ..., N\} - \{i\}$, the balance of token $j$ after the withdrawal is $B_j$. Thus, the value of Balancer's value function after the single-asset withdrawal is

$$V'_1 = \left(B_i - a'_i\right)^{W_i} \prod_{\substack{j=1 \\ j \neq i}}^{N} B_j^{W_j}.$$

# 3.5 Pool Tokens Swap

Consider two different Balancer pools, $P_1$ and $P_2$. Let $BPT_1$ and $BPT_2$ be the BPT of pools $P_1$ and $P_2$, respectively. Let $M_1$ be the total supply of $BPT_1$ and let $M_2$ be the total supply of $BPT_2$. We want to find the fair exchange price between $BPT_1$ and $BPT_2$. We will proceed as follows. We will choose a token $X_1$ of pool $P_1$ and a token $X_2$ of pool $P_2$ and compute the values of pools $P_1$ and $P_2$ in terms of tokens $X_1$ and $X_2$, respectively. Then we will choose another token $Z$ and compute the values of pools $P_1$ and $P_2$ in terms of token $Z$ in order to find the value of each $BPT_1$ token and each $BPT_2$ token in terms of token $Z$.

Let $X_1$ be a token of pool $P_1$ and let $X_2$ be a token of pool $P_2$. For each $j \in \{1, 2\}$, let $B_j$ be the balance of token $X_j$ in pool $P_j$ and let $W_j$ be the weight of token $X_j$ in pool $P_j$. Let $Z$ be another token, and for $j \in \{1, 2\}$, let $p_j$ be the price of token $X_j$ in terms of token $Z$.

From Equation 3.2, we obtain that for each $j \in \{1, 2\}$, the value of the whole pool $P_j$ in terms of token $X_j$ is $\dfrac{B_j}{W_j}$. Then, for all $j \in \{1, 2\}$, the value of each $BPT_j$ token in terms of token $X_j$ is

$$V_{BPT_j}\left(X_j\right) = \frac{B_j}{W_j M_j}$$

and hence, the value of each BPT$_j$ token in terms of token $Z$ is

$$V_{\mathrm{BPT}_j}(Z) = p_j \cdot \frac{B_j}{W_j M_j}$$

Let $P$ be the price of one BPT$_1$ token in terms of BPT$_2$ tokens. Hence, $V_{\mathrm{BPT}_1}(Z) = P \cdot V_{\mathrm{BPT}_2}(Z)$, and thus,

$$P = \frac{V_{\mathrm{BPT}_1}(Z)}{V_{\mathrm{BPT}_2}(Z)} = \frac{B_1}{B_2} \cdot \frac{W_2 M_2}{W_1 M_1} \cdot \frac{p_1}{p_2}. \qquad (3.10)$$

Summing up, this means that each BPT$_1$ token is worth

$$\frac{B_1}{B_2} \cdot \frac{W_2 M_2}{W_1 M_1} \cdot \frac{p_1}{p_2}$$

BPT$_2$ tokens.

**Example 3.9.** Suppose that Alice and Bob are liquidity providers of two different Balancer pools described by the data given in the following table:

|  | Alice's pool | Bob's pool |
| --- | --- | --- |
| BPT supply | 10,000 | 20,000 |
| Token X | ETH | ETH |
| Token Y | USDC | USDC |
| Balance of token X | 9,000 | 6,000 |
| Balance of token Y | 9,000,000 | 16,000,00 |
| Weight of token X | 0.8 | 0.6 |
| Weight of token Y | 0.2 | 0.4 |

We will apply Equation 3.10 using ETH as token $Z$. Hence, the price of one of Alice's BPT in terms of Bob's BPT is

$$P = \frac{B_1}{B_2} \cdot \frac{W_2 M_2}{W_1 M_1} \cdot \frac{p_1}{p_2} = \frac{9,000}{6,000} \cdot \frac{0.6 \cdot 20,000}{0.8 \cdot 10,000} \cdot \frac{1}{1} = 2.25$$

This means that if Alice gives Bob 1 BPT from her pool, she will get 2.25 BPT of Bob's pool in return.

Observe also that applying Equation 3.2, we obtain that the total value of Alice's pool is 45,000,000 USDC and the total value of Bob's pool is 40,000,000 USDC. Hence, each of Alice's BPT is worth $\dfrac{45,000,000}{10,000} = 4,500$ USDC, while each of Bob's BPT is worth $\dfrac{40,000,000}{20,000} = 2,000$ USDC. From this, we can see that each of Alice's BPT is worth 2.25 of Bob's BPT.

## 3.5.1 Swap of Pool Tokens That Belong to Pools of Different Types

We will now analyze an example in which we would like to swap BPT of a Balancer pool for LP tokens of a Uniswap v2 pool. Recall that the Uniswap v2 AMM can be regarded as a particular case of the Balancer AMM, so the procedure for swapping BPT that we explained before can be applied in this case.

**Example 3.10.** Suppose that Alice is a liquidity provider of an unequally weighted Balancer pool and Bob is a liquidity provider of a Uniswap v2 pool, or equivalently, a Balancer pool having two equally weighted tokens. Assume that Alice's pool and Bob's pool are described in the following table:

|  | Alice's pool (Balancer) | Bob's pool (Uniswap v2) |
|---|---|---|
| Supply of pool tokens | 10,000 | 20,000 |
| Token X | ETH | ETH |
| Token Y | USDC | USDC |
| Balance of token X | 3,600 | 3,000 |
| Balance of token Y | 3,600,000 | 12,000,000 |
| Weight of token X | 0.8 | 0.5 |
| Weight of token Y | 0.2 | 0.5 |

Again, we will apply Equation 3.10 using ETH as token $Z$. Hence, the price of one of Alice's BPT in terms of Bob's BPT is

$$P = \frac{B_1}{B_2} \cdot \frac{W_2 M_2}{W_1 M_1} \cdot \frac{p_1}{p_2} = \frac{3,600}{3,000} \cdot \frac{0.5 \cdot 20,000}{0.8 \cdot 10,000} \cdot \frac{1}{1} = 1.5$$

This means that if Alice gives Bob 1 BPT from her pool, she will get 1.5 LP tokens of Bob's pool in return.

Observe also that applying Equation 3.2, we obtain that the total value of Alice's pool is 18,000,000 USDC and the total value of Bob's pool is 24,000,000 USDC. Hence, each of Alice's BPT is worth $\frac{18,000,000}{10,000} = 1,800$ USDC, while each of Bob's LP tokens is worth $\frac{24,000,000}{20,000} = 1,200$ USDC. From this, we can see that each of Alice's BPT is worth 1.5 of Bob's LP tokens.

# 3.6 Summary

In this chapter, we analysed the Balancer AMM and gave the corresponding spot price and trading formulae. We also explained how all-asset and single-asset deposits and withdrawals work and gave interesting mathematical results comparing all-asset deposits with single-asset deposits.

In the next chapter, we will study the StableSwap AMM, which is designed to allow trades between pegged assets with low price impact.

# CHAPTER 4

# Curve Finance

Curve Finance's AMM was originally designed to facilitate trades with low price impact between stablecoins or, more generally, between assets that have the same price–like ETH and sETH (Synthetix[1] ETH). Curve's AMM was launched in January 2020 and was initially called StableSwap. Its algorithm is more complex than those of the AMMs we have studied in the previous chapters.

In June 2021, Curve v2 was presented [12]. This new version of Curve's AMM is called Crypto Pools, and although it is based on the previous one, it introduces a new algorithm to allow trading between nonpegged assets.[2]

Currently, both versions of the Curve AMM coexist. In this chapter, we will explain how the StableSwap AMM works and analyze the maths behind it. We will not cover Curve v2 in this book.

## 4.1 The StableSwap Invariant

The StableSwap invariant is a combination of the constant product invariant and the constant sum invariant, which are defined as

$$\prod_{j=1}^{N} b_j = C \ \text{ and } \ \sum_{j=1}^{N} b_j = D$$

---

[1] https://synthetix.io/
[2] https://curve.fi/

© Miguel Ottina, Peter Johannes Steffensen, Jesper Kristensen 2023
M. Ottina et al., *Automated Market Makers*, https://doi.org/10.1007/978-1-4842-8616-6_4

respectively, where $C$ and $D$ are positive real numbers, $N$ is the number of tokens in the pool, and, for each $j \in \{1, 2, ..., N\}$, $b_j$ denotes the balance of token $j$ in the pool. Observe that the product invariant is equivalent to Balancer's geometric mean invariant with equally weighted tokens. On the other hand, note that the constant sum invariant gives a constant price of 1 for each pair of tokens, since if we deposit an amount $b$ of token $i$ and receive an amount $a$ of token $o$, from the constant sum invariant, we obtain that

$$\sum_{j=1}^{N} b_j = (b_i + b) + (b_o - a) + \sum_{\substack{j=1 \\ j \neq i, o}}^{N} b_j,$$

and hence, $b_i + b_o = b_i + b + b_o - a$, from where it follows that $a = b$, which implies that the effective price paid per unit of token $o$ in terms of token $i$ is $\frac{b}{a} = 1$. In particular, there is no difference between the spot price and the effective price paid by a trader; that is, there is no price impact.

As we previously mentioned, the StableSwap AMM invariant is obtained as a combination of the constant sum and constant product invariants. We will explain now how it is constructed. Let $N$ be the number of tokens in the pool, and for $j \in \{1, 2, ..., N\}$, let $b_j$ be a variable that represents the balance of token $j$ in the pool. Recall that the constant sum invariant is given by

$$\sum_{j=1}^{N} b_j = D$$

for a certain positive number $D$ and the constant product invariant is given by

$$\prod_{j=1}^{N} b_j = C$$

for a certain positive number $C$. In order to combine these invariants, we will establish a relation between the numbers $C$ and $D$. To this end, assume for a moment that all tokens are equally priced. Hence, for all $j$, $k \in \{1, 2, ..., N\}$, the spot price of token $j$ in terms of token $k$ is 1, and then from Equation 3.1, we obtain that $b_j = b_k$ for all $j, k \in \{1, 2, ..., N\}$. Thus, from the constant sum formula, we obtain that $b_j = \dfrac{D}{N}$ for all $j \in \{1, 2, ..., N\}$, and substituting these expressions in the constant product formula, we obtain that $C = \left(\dfrac{D}{N}\right)^N$. Thus, our constant product invariant will be

$$\prod_{j=1}^{N} b_j = \left(\frac{D}{N}\right)^N.$$

The first idea to combine both invariants is to take a convex combination of both formulae; that is, for any chosen $t \in [0, 1]$, we consider the expression

$$t\sum_{j=1}^{N} b_j + (1-t)\prod_{j=1}^{N} b_j = tD + (1-t)\left(\frac{D}{N}\right)^N.$$

Note that when $t = 0$, we obtain the constant product formula, and when $t = 1$, we obtain the constant sum formula.

Now we will modify the previous expression, replacing the parameter $t$ with a new parameter $x$ so that we obtain the constant product formula when $x = 0$ and the constant sum formula when $x$ tends to infinity. To this end, we need to replace the parameter $t$ with a function $g(x)$ such that

$$g(0) = 0 \text{ and } \lim_{x \to +\infty} g(x) = 1.$$

Although many functions can do the trick, we consider the map $g : [0, +\infty) \to \mathbb{R}$ given by $g(x) = \dfrac{x}{x+1}$. Note that this function $g$ satisfies the desired properties and that $0 \leq g(x) \leq 1$ for all $x \geq 0$. Replacing $t = g(x)$ in the preceding expression, we obtain

$$\frac{x}{x+1}\sum_{j=1}^{N}b_j + \frac{1}{x+1}\prod_{j=1}^{N}b_j = \frac{x}{x+1}D + \frac{1}{x+1}\left(\frac{D}{N}\right)^N$$

or equivalently,

$$x\sum_{j=1}^{N}b_j + \prod_{j=1}^{N}b_j = xD + \left(\frac{D}{N}\right)^N.$$

In addition, we want the parameter $x$ to be dimensionless. To this end, we will substitute $x = \chi D^{N-1}$ to obtain

$$\chi D^{N-1}\sum_{j=1}^{N}b_j + \prod_{j=1}^{N}b_j = \chi D^N + \left(\frac{D}{N}\right)^N.$$

Finally, we want the parameter $\chi$ to be dynamic rather than a constant chosen value. To this end, we substitute

$$\chi = A \cdot \frac{\displaystyle\prod_{j=1}^{N}b_j}{\left(\dfrac{D}{N}\right)^N}, \qquad (4.1)$$

where $A$ is a positive number that is called *amplification coefficient*. In the actual code, there is a parameter labeled A, which is a positive integer and is equal to $AN^{N-1}$. Therefore, from now on, we will assume that $AN^{N-1}$ is a positive integer–and this will indeed be needed later.

The idea behind the substitution of Equation 4.1 is the following. Assuming that $\sum_{j=1}^{N}b_j = D$ (which will not hold in general as the pool balances change), by the inequality of arithmetic and geometric means,[3] we have that

---

[3] See, for example, [10, Theorem 5.1] or [26, Theorem 1.3.1].

$$\left(\prod_{j=1}^{N} b_j\right)^{\frac{1}{N}} \leq \frac{1}{N}\sum_{j=1}^{N} b_j = \frac{D}{N},$$

(with equality holding if and only if $b_1 = b_2 = \ldots = b_N$). Hence,

$$\prod_{j=1}^{N} b_j \leq \left(\frac{D}{N}\right)^{N},$$

and thus, $\chi \leq A$. On the other hand, when the balances move away of the equilibrium point $b_1 = b_2 = \ldots = b_N$, the parameter $\chi$ decreases, as we can see for the case $N = 2$ in Figure 4-1, where we plot the graph of the function $f: [0, 2] \times [0, 2] \to \mathbb{R}$ given by

$$f(b_1, b_2) = \frac{b_1 b_2}{\left(\dfrac{b_1 + b_2}{2}\right)^{2}}.$$

As we can see, the function has maximum value 1 when $b_1 = b_2$ and decreases when we move away from this line.

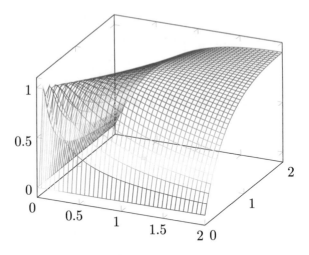

**Figure 4-1.** *Graph of the function f*

Making the aforementioned substitution of $\chi$, we obtain

$$A \cdot \frac{\prod_{j=1}^{N} b_j}{\left(\dfrac{D}{N}\right)^N} D^{N-1} \sum_{j=1}^{N} b_j + \prod_{j=1}^{N} b_j = A \cdot \frac{\prod_{j=1}^{N} b_j}{\left(\dfrac{D}{N}\right)^N} D^N + \left(\frac{D}{N}\right)^N,$$

and multiplying both sides by $\dfrac{D}{N \prod\limits_{j=1}^{N} b_j}$ , we get

$$AN^N \sum_{j=1}^{N} b_j + D = ADN^N + \frac{D^{N+1}}{N^N \prod\limits_{j=1}^{N} b_j}, \tag{4.2}$$

which is the StableSwap invariant [11].

In Figure 4-2, we compare the StableSwap invariant for three different values of the amplification coefficient $A$ with the constant product and constant sum invariants in a pool with two tokens. Observe that at the equilibrium point, the price of token $X$ in terms of token $Y$ is 1, since the slope of the curve is –1. In addition, near the equilibrium point, the curve looks like a constant sum invariant curve, and hence, the price impact is very low, and the price is approximately 1. But if the balances of the tokens deviate from the equilibrium point, the price changes, and the curve looks more like a constant product invariant one.

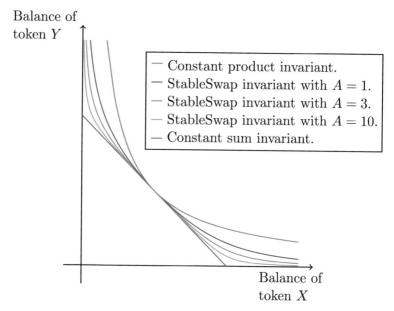

**Figure 4-2.** *Comparison between the StableSwap invariant, the constant product invariant, and the constant sum invariant*

# 4.2 Mathematical Preliminaries

In this subsection, we will analyze several mathematical arguments that are used within the StableSwap AMM. We will also prove some results that will be needed later in order to explain the mathematical foundations of this AMM.

Consider a StableSwap liquidity pool consisting of $N$ tokens. For each $j \in \{1, 2, ..., N\}$, let $B_j$ be the balance of token $j$ in the pool. Let $D_0$ be the value of the pool parameter $D$. Recall that $D_0 > 0$. Let $o \in \{1, 2, ..., N\}$. We will prove that the balance of token $o$ is uniquely determined by $D_0$ and the balances $B_j, j \neq o$.

For simplicity, let $y = B_o$. From the invariant Equation 4.2, we know that

$$AN^N \sum_{j=1}^{N} B_j + D_0 = AD_0 N^N + \frac{D_0^{N+1}}{N^N \prod\limits_{j=1}^{N} B_j},$$

that is,

$$AN^N y + AN^N \sum_{\substack{j=1 \\ j \neq o}}^{N} B_j + D_0 - AD_0 N^N = \frac{D_0^{N+1}}{N^N y \prod\limits_{\substack{j=1 \\ j \neq o}}^{N} B_j}.$$

Multiplying both sides of the equation by $\dfrac{y}{AN^N}$, we get

$$y^2 + \left( \sum_{\substack{j=1 \\ j \neq o}}^{N} B_j + \frac{D_0}{AN^N} - D_0 \right) y = \frac{D_0^{N+1}}{AN^{2N} \prod\limits_{\substack{j=1 \\ j \neq o}}^{N} B_j}.$$

Let

$$S = \sum_{\substack{j=1 \\ j \neq o}}^{N} B_j, \qquad\qquad P = \prod_{\substack{j=1 \\ j \neq o}}^{N} B_j,$$

$$b = S + \frac{D_0}{AN^N}, \text{ and } c = \frac{D_0^{N+1}}{AN^{2N} P}.$$

With this notations, the previous equation can be written as

$$y^2 + (b - D_0)y - c = 0, \qquad (4.3)$$

and the value of $y$ can be obtained through solving the previous quadratic equation. Observe that if $y_1$ and $y_2$ are the solutions of the previous equation, then $y_1 y_2 = -c$, and since $c > 0$, we obtain that $y_1 y_2 < 0$. This means that one of the solutions of the previous equation is positive and the other is negative. Thus, the value of $y$ that we are looking for is well defined since it is the only positive solution to the previous equation. Therefore, we have proved that the balance of token $o$ is uniquely determined by $D_0$ and the balances $B_j, j \neq 0$.

The previous proof allows us to give the following definition.

**Definition 4.1.** Consider a StableSwap liquidity pool that has $N$ tokens. Let $o \in \{1, 2, ..., N\}$. For each $j \in \{1, 2, ..., N\} - \{o\}$, let $B_j$ be the balance of token $j$ in the pool. Let $D_0$ be the value of the pool parameter $D$. We define

$$\mathcal{Y}(o, B_1, ..., B_{o-1}, B_{o+1}, ..., B_N, D_0)$$

as the only positive number that satisfies Equation 4.3. That is, the number $\mathcal{Y}(o, B_1, ..., B_{o-1}, B_{o+1}, ..., B_N, D_0)$ is the balance of token $o$ if the balance of token $j$ is $B_j$ for all $j \in \{1, 2, ..., N\} - \{o\}$ and the value of the pool parameter $D$ is $D_0$.

We will analyze now how $\mathcal{Y}(o, B_1, ..., B_{o-1}, B_{o+1}, ..., B_N, D_0)$ changes when we modify the balances $B_j, j \neq o$, or the value of the pool parameter $D$.

**Proposition 4.1.** *Consider a StableSwap liquidity pool of N tokens. Let $o \in \{1, 2, ..., N\}$. For each $j \in \{1, 2, ..., N\} - \{o\}$, let $B_j$ and $B'_j$ be positive real numbers. Let*

$$\mathbf{B} = (B_1, ..., B_{o-1}, B_{o+1}, ..., B_N)$$

*and let*

$$\mathbf{B}' = \left( B'_1, \ldots, B'_{o-1}, B'_{o+1}, \ldots, B'_N \right).$$

*Let $D_0$ and $D'_0$ be positive real numbers. If $D_0 \geq D'_0$ and $B_j \leq B'_j$ for all $j \in \{1, 2, \ldots, N\} - \{o\}$, then*

$$\mathcal{Y}(o, \mathbf{B}, D_0) \geq \mathcal{Y}(o, \mathbf{B}', D'_0).$$

*If, in addition, any of the inequalities $B_j \leq B'_j, j \neq o$, or $D_0 \geq D'_0$ is strict, then*

$$\mathcal{Y}(o, \mathbf{B}, D_0) > \mathcal{Y}(o, \mathbf{B}', D'_0).$$

*Proof. Let*

$$y = \mathcal{Y}(o, \mathbf{B}, D_0)$$

and let

$$y' = \mathcal{Y}(o, \mathbf{B}', D'_0).$$

Let

$$S = \sum_{\substack{j=1 \\ j \neq o}}^{N} B_j, \quad S' = \sum_{\substack{j=1 \\ j \neq o}}^{N} B'_j,$$

$$P = \prod_{\substack{j=1 \\ j \neq o}}^{N} B_j, \quad P' = \prod_{\substack{j=1 \\ j \neq o}}^{N} B'_j,$$

$$b = S + \frac{D_0}{AN^N}, \quad b' = S' + \frac{D'_0}{AN^N},$$

$$c = \frac{D_0^{N+1}}{AN^{2N}P}, \text{ and } c' = \frac{\left(D_0'\right)^{N+1}}{AN^{2N}P'}.$$

Recall that $y^2 + (b - D_0)y - c = 0$ and $\left(y'\right)^2 + \left(b' - D_0'\right)y' - c' = 0$ and that $y > 0$ and $y' > 0$. Hence,

$$y + \left(b - D_0\right) = \frac{c}{y} \text{ and } y' + \left(b' - D_0'\right) = \frac{c'}{y'}.$$

Suppose that $B_j \leq B_j'$ for all $j \in \{1, 2, ..., N\} - \{o\}$ and $D_0 \geq D_0'$. Then $S \leq S'$ and $P \leq P'$. And since $D_0 \geq D_0'$, we obtain that $c \geq c'$. On the other hand, since $AN^{N-1}$ is a positive integer and $N \geq 2$, it follows that $\frac{1}{AN^N} - 1 < 0$. Thus,

$$b - D_0 = S + \frac{D_0}{AN^N} - D_0 = S + D_0\left(\frac{1}{AN^N} - 1\right)$$

$$\leq S' + D_0'\left(\frac{1}{AN^N} - 1\right) = S' + \frac{D_0'}{AN^N} - D_0'$$

$$= b' - D_0'$$

Suppose that $y < y'$. Then

$$\frac{c}{y} > \frac{c'}{y'} = y' + \left(b' - D_0'\right) > y + \left(b - D_0\right) = \frac{c}{y}$$

which entails a contradiction. Thus, $y \geq y'$.

In addition, if any of the inequalities $B_j \leq B_j', j \neq o$, or $D_0 \geq D_0'$ is strict, then $c > c'$. Suppose that $y \leq y'$. Then

$$\frac{c}{y} > \frac{c'}{y'} = y' + \left(b' - D_0'\right) \geq y + \left(b - D_0\right) = \frac{c}{y}$$

which entails a contradiction. Thus, $y > y'$. $\qquad \Box$

## 4.2.1 Finding the Parameter D

After each transaction, the pool parameter $D$ is updated based on the current pool state. Given a liquidity pool with $N$ tokens and balances $B_1, B_2, ..., B_N$, the parameter $D$ can be found by solving the following equation:

$$\frac{D^{N+1}}{N^N \prod\limits_{j=1}^{N} B_j} + D\left(AN^N - 1\right) - AN^N \sum_{j=1}^{N} B_j = 0 \qquad (4.4)$$

which follows from the invariant Equation 4.2.

Note that in order for the value $D$ to be well defined, we need the previous equation to have a unique positive solution. We will prove now that this is indeed the case. Let $g : \mathbb{R} \to \mathbb{R}$ be defined by

$$g(x) = \frac{1}{N^N \prod\limits_{j=1}^{N} B_j} x^{N+1} + \left(AN^N - 1\right)x - AN^N \sum_{j=1}^{N} B_j.$$

Note that $g$ is a polynomial of degree $N + 1$ (and hence a continuous function) and that $AN^N - 1 > 0$ since $AN^{N-1}$ is a positive integer and $N \geq 2$. On the other hand, the derivative of $g$ is

$$g'(x) = \frac{N+1}{N^N \prod\limits_{j=1}^{N} B_j} x^N + \left(AN^N - 1\right).$$

Observe that $g'(x) > 0$ for all $x \geq 0$ since $N, B_1, B_2, ..., B_N$ and $AN^N - 1$ are positive numbers. Thus, $g$ is a strictly increasing function in the interval $[0, +\infty)$. It follows that $g$ has at most one root in that interval. In addition, since

$$g(0) = -AN^N \sum_{j=1}^{N} B_j < 0 \quad \text{and} \quad \lim_{x \to +\infty} g(x) = +\infty$$

(because the leading coefficient of $g$ is a positive number), we obtain that $g$ has at least one root in the interval $(0, +\infty)$. Therefore, $g$ has exactly one positive root, and thus, the parameter $D$ is well defined.

Now, in order to effectively compute the value of the parameter $D$, the smart contract applies the Newton's Method[4] to the function $g$. Explicitly, let $d_0 > 0$ be an initial approximate value for the parameter $D$. For example, we can take

$$d_0 = \sum_{j=1}^{N} B_j,$$

where $B_1, B_2, \ldots, B_N$ are the balances of the tokens of the pool. Now let

$$d_1 = d_0 - \frac{g(d_0)}{g'(d_0)}.$$

Then, for each $n \in \mathbb{N}$, let

$$d_{n+1} = d_n - \frac{g(d_n)}{g'(d_n)}. \tag{4.5}$$

The process ends when the desired precision has been obtained, that is, when the difference between two consecutive terms, $d_k$ and $d_{k+1}$, is smaller than a certain number.

This process can be found in the get $D$ function of the smart contract of the StableSwap AMM,[5] as we will see now. For simplicity, let

$$\alpha = AN^N - 1 \quad \text{and} \quad \beta = \frac{1}{N^N \prod_{j=1}^{N} B_j}.$$

---

[4] See, for example, [24, Section 3.8].
[5] The smart contracts of the different StableSwap pools can be found in
https://github.com/curvefi/curve-contract

Note that

$$d_{n+1} = d_n - \frac{g(d_n)}{g'(d_n)} = d_n - \frac{\beta d_n^{N+1} + \alpha d_n - AN^N \sum_{j=1}^{N} B_j}{(N+1)\beta d_n^N + \alpha}$$

$$= \frac{(N+1)\beta d_n^{N+1} + \alpha d_n - \beta d_n^{N+1} - \alpha d_n + AN^N \sum_{j=1}^{N} B_j}{(N+1)\beta d_n^N + \alpha}$$

$$= \frac{N\beta d_n^{N+1} + AN^N \sum_{j=1}^{N} B_j}{(N+1)\beta d_n^N + \alpha} = \frac{\left(N\beta d_n^{N+1} + AN^N \sum_{j=1}^{N} B_j\right) d_n}{(N+1)\beta d_n^{N+1} + \alpha d_n}.$$

This formula can be found in the code shown in Listing 4-1, where the translation between the variables of the code and our notation is the following:

$$\textbf{N\_COINS} = N,$$
$$\textbf{Ann} \quad = AN^N,$$
$$\textbf{S} \quad = \sum_{j=1}^{N} B_j,$$
$$\textbf{D} \quad = d_n,$$
$$\textbf{D\_P} \quad = \beta d_n^{N+1} = \frac{d_n^{N+1}}{N^N \prod_{j=1}^{N} B_j}.$$

Note also that $d_0$ = S.

***Listing 4-1.*** get_D function of the smart contract of StableSwap

```
def get_D (xp: uint256[N_COINS], amp: uint256) ->
↪ uint256:
    S: uint256 =0
    for _x in xp:
        S += _x
    if S == 0:
        return 0

    Dprev: uint256 = 0
    D: uint256 = S
    Ann: uint256 = amp * N_COINS
    for _i in range(255):
        D_P: uint256 = D
        for _x in xp:
            D_P = D_P * D / (_x * N_COINS) # If
            ↪    division by 0, this will be borked:
            ↪    only withdrawal will work. And that is
            ↪    good
        Dprev = D
        D = (Ann * S + D_P * N_COINS) * D / ((Ann - 1)
        ↪    * D + (N_COINS + 1) * D_P)
        # Equality with the precision of 1
        if D > Dprev:
            if D - Dprev <= 1:
                break
        else:
            if Dprev - D <= 1:
                break
    return D
```

# 4.3 Trading Formulae

Consider a StableSwap liquidity pool of $N$ tokens and no fees and let $B_1$, $B_2, ..., B_N$ be the balances of the tokens in the pool. Let $D_0$ be the value of the pool parameter $D$. Let $i, o \in \{1, 2, ..., N\}$. As in the previous chapters, we want to compute the amount $A_o$ of token $o$ that a trader will receive when depositing a certain amount $A_i$ of token $i$. For each $j \in \{1, 2, ..., N\}$, let $B_j'$ be the balance of token $j$ after the trade. Clearly, $B_o' = B_o - A_o$, $B_i' = B_i + A_i$ and $B_j' = B_j$ for all $j \in \{1, 2, ..., N\} - \{i, o\}$. For simplicity, let $y = B_o'$. Note that $A_o = B_o - y$. We will compute $y$. From the previous section, we know that the value of $y$ can be obtained by solving the quadratic Equation 4.3 and is given by $y = \mathcal{Y}(o, B_1', ..., B_{o-1}', B_{o+1}', ..., B_N', D_0)$ following Definition 4.1.

In the code of the StableSwap AMM, instead of solving the quadratic equation, a different approach is taken: the value of $y$ is found using Newton's Method, as we can see in the get_y function shown in Listing 4-2.

*Listing 4-2.* get_y function of the smart contract of StableSwap

```
def get_y(i: int128, j: int128, x: uint256, xp_:
↪  uint256[N_COINS]) -> uint256:
    # x in the input is converted to the same
    ↪  price/precision

    assert i != j       # dev: same coin
    assert j >= 0       # dev: j below zero
    assert j < N_COINS  # dev: j above N_COINS

    # should be unreachable, but good for safety
    assert i >= 0
    assert i < N_COINS

    amp: uint256 = self._A()
    D: uint256 = self.get_D(xp_, amp)
    c: uint256 = D
```

```
    S_: uint256 = 0
    Ann: uint256 = amp * N_COINS

    _x: uint256 = 0
    for _i in range(N_COINS):
        if _i == i:
            _x = x
        elif _i != j:
            _x = xp_[_i]
        else:
            continue
        S_ += _x
        c = c * D / (_x * N_COINS)
c = c * D / (Ann * N_COINS)
b: uint256 = S_ + D / Ann # - D
y_prev: uint256 = 0
y: uint256 = D
for _i in range(255):
    y_prev = y
    y = (y*y + c) / (2 * y + b - D)
    # Equality with the precision of 1
    if y > y_prev:
        if y - y_prev <= 1:
            break
    else:
        if y_prev - y <= 1:
            break
return y
```

## 4.3.1 Taking Fee into Consideration

Unlike Uniswap and Balancer, the pool fee for the StableSwap AMM is charged on the output token. In consequence, with the previous notations, after finding the balance $y$ of token $o$ using Equation 4.3, the amount $A_o$ of token $o$ that the trader receives is

$$A_o = (B_o - y) \cdot (1 - \phi)$$

where $\phi$ is the pool fee. Note that the trader is charged a fee of $\phi(B_o - y)$. In addition, a fraction $\omega$ of this charged fee is generally reserved for the protocol as an administrative fee and not deposited into the pool. And since, if there had been no fees, the balance of token $o$ after the trade would have been equal to $y$, the balance of token $o$ is updated in the following way:

$$B_0' = y + (B_o - y)\phi - \omega(B_o - y)\phi = y + (B_o - y)\phi(1 - \omega).$$

Observe that

$$B_0' = y + (B_o - y)\phi - \omega(B_o - y)\phi$$

$$= B_o - (B_o - y) + (B_o - y)\phi - \omega(B_o - y)\phi$$

$$= B_o - (B_o - y)(1 - \phi) - \omega(B_o - y)\phi$$

$$= B_o - A_o - \omega(B_o - y)\phi.$$

This last formula is the one that appears in the code of the StableSwap AMM, as we can see in Listing 4-3.

On the other hand, the updated balance of token $i$ is clearly

$$B'_i = B_i + A_i.$$

**Listing 4-3.** exchange function of the smart contract of StableSwap

```
def exchange(i: int128, j: int128, dx: uint256, min_dy:
↪  uint256):
    assert not self.is_killed # dev: is killed
    rates: uint256[N_COINS] = RATES

    old_balances: uint256[N_COINS] = self.balances
    xp: uint256[N_COINS] = self._xp_mem(old_balances)

    # Handling an unexpected charge of a fee on
    ↪  transfer (USDT, PAXG)
    dx_w_fee: uint256 = dx
    input_coin: address = self.coins[i]

    if i == FEE_INDEX:
        dx_w_fee = ERC20(input_coin).balanceOf(self)

    # "safeTransferFrom" which works for ERC20s which
    ↪  return bool or not
    _response: Bytes[32] = raw_call(
        input_coin,
        concat(

            ↪  method_id("transferFrom(address,address,uint256)")
            convert(msg.sender, bytes32),
            convert(self, bytes32),
            convert(dx, bytes32),
        ),
    max_outsize=32,
```

```
) # dev: failed transfer
if len(_response) > 0:
    assert convert(_response, bool) # dev: failed
    ↳  transfer

if i == FEE_INDEX:
    dx_w_fee = ERC20(input_coin).balanceOf(self) -
    ↳  dx_w_fee

x: uint256 = xp[i] + dx_w_fee * rates[i] /
↳  PRECISION
y: uint256 = self.get_y(i, j, x, xp)

dy: uint256 = xp[j] - y - 1 # -1 just in case
↳  there were some rounding errors
dy_fee: uint256 = dy * self.fee / FEE_DENOMINATOR

# Convert all to real units
dy = (dy - dy_fee) * PRECISION / rates[j]
assert dy >= min_dy, "Exchange resulted in fewer
↳  coins than expected"
dy_admin_fee: uint256 = dy_fee * self.admin_fee /
↳  FEE_DENOMINATOR
dy_admin_fee = dy_admin_fee * PRECISION / rates[j]

# Change balances exactly in same way as we change
↳  actual ERC20 coin amounts
self.balances[i] = old_balances[i] + dx_w_fee
# When rounding errors happen, we undercharge admin
↳  fee in favor of LP
self.balances[j] = old_balances[j] - dy -
↳  dy_admin_fee

# "safeTransfer" which works for ERC20s which
```

```
↪   return bool or not
_response = raw_call(
    self.coins[j],
    concat(
        method_id("transfer(address,uint256)"),
        convert(msg.sender, bytes32),
        convert(dy, bytes32),
    ),
    max_outsize=32,
)   # dev: failed transfer
if len(_response) > 0:
    assert convert(_response, bool) # dev: failed
    ↪   transfer

log TokenExchange(msg.sender, i, dx, j, dy)
```

**Example 4.1.** Consider a StableSwap pool having three tokens, USDC, DAI, and USDT, with the following balances:

|       | Token 1 | Token 2 | Token 3 |
|-------|---------|---------|---------|
|       | USDC    | DAI     | USDT    |
| $B_j$ | 100,000 | 120,000 | 80,000  |

Suppose that the fee of the pool is 0.03%, that the pool administrative fee is 50%, and that the value of the amplification coefficient $A$ is 500.

Applying the algorithm described by Equation 4.5, we obtain that the value of $D$ is approximately 299,999. Observe that it is very close to the sum $B_1 + B_2 + B_3$ of the balances of each token, which is 100,000+120,000+80,000=300,000.

Suppose that a trader wants to obtain USDC depositing 10,000 USDT. In order to compute the amount of USDC that the trader receives, we first compute the value of $y$ from Equation 4.3, which represents the balance of USDC in the pool after the deposit of 10,000 USDT, keeping constant the value of $D$ and assuming that there are no fees. Solving the quadratic equation, we obtain that $y \approx 89{,}999.71$. Hence, the amount $A_o$ of USDC that the trader receives is

$$A_o = (B_o - y) \cdot (1 - \phi) = (100{,}000 - 89{,}999.71) \cdot 0.9997$$
$$\approx 9{,}997.29$$

The updated balance of USDC in the pool is given by

$$B_0' = B_o - A_o - \omega(B_o - y)\phi$$

$$\approx 100{,}000 - 9{,}997.29 - 0.5 \cdot (100{,}000 - 89{,}999.71) \cdot 0.0003$$

$$\approx 90{,}001.21$$

Observe that the balance of USDC in the pool has decreased by approximately 100,000 - 90,001.21 = 9,998.79, and since just 9,997.29 USDC are given to the trader, we obtain that the difference between those amounts–1.5 USDC–has been charged as an administrative fee to the liquidity providers, since an additional amount of 1.5 USDC has been removed from the pool.

Note also that if there had been no fees, the trader would have received approximately 100,000 – 89,999.71 = 10,000.29 USDC (this is 3 USDC more than the amount they received when fees were considered), and thus, the balance of USDC in the pool would have been 100,000 - 10,000.29 = 89,999.71 (which is represented by $y$). Therefore, the pool now has 90,001.21 - 89,999.71 = 1.5 USDC more than it would have had if there were no fees. As we can see, the trader has been charged a fee of 3 USDC, and from this amount, 1.5 USDC goes into the pool, and the other 1.5 USDC is reserved separately as an administrative fee of the protocol.

Finally, note that the balances of the tokens in the pool after the trade are as follows:

|  | USDC | DAI | USDT |
|---|---|---|---|
| $B_j$ | 90,001.21 | 120,000 | 90,000 |

Also, after the trade, the parameter $D$ is recalculated with the algorithm given by Equation 4.5, and we have that the new value for the parameter $D$ is approximately 300,000.57.

# 4.4 All-Asset Deposit

In a similar way as with Uniswap and Balancer, liquidity providers can deposit assets into a StableSwap pool in order to earn trading fees. In exchange for their deposit, they receive StableSwap LP tokens, which represent ownership of the assets contained in the pool. During the pool trading activity, the fees from swaps accrue into the pool, resulting in added value for the StableSwap LP tokens. In this section, we will describe how the all-asset deposit works in the StableSwap AMM.

Consider a StableSwap AMM with $N$ tokens, and suppose that a liquidity provider wants to deposit amounts $x_1, x_2, ..., x_N$ of tokens 1, 2, ..., $N$, respectively. Let $B_1, B_2, ..., B_N$ be the pool balances before the deposit. Since the pool needs to remain balanced, the amounts $x_1, x_2, ..., x_N$ must satisfy that

$$\frac{x_i}{B_i} = \frac{x_j}{B_j}$$

for all $i, j \in \{1, 2, ..., N\}$.

Let $M$ be the amount of existent StableSwap LP tokens before the deposit and let $q = \frac{x_1}{B_1}$ be the share that the new liquidity provider adds to the pool. Hence, they will receive a share $q$ of the amount of existent StableSwap LP tokens; that is, they will receive an amount $qM$ of StableSwap LP tokens. Note also that for all $j \in \{1, 2, ..., N\}$,

$$\frac{x_j}{B_j} = \frac{x_1}{B_1} = q,$$

and thus, $x_j = qB_j$ for all $j \in \{1, 2, ..., N\}$. Hence, in order to receive a share $q$ of the existent StableSwap LP tokens, the liquidity provider has to deposit, for each $j \in \{1, 2, ..., N\}$, an amount $qB_j$ of token $j$.

We will prove now that the number of newly minted StableSwap LP tokens is exactly reflected in the change of the StableSwap pool parameter $D$. Concretely, assume that the new liquidity provider deposits a share $q$ of each token. Let $D_0$ and $D_1$ be the values of the pool parameter $D$ before and after the deposit, respectively. Let $B_1, B_2, ..., B_N$ be the balances of the tokens in the pool before the deposit, and let $B_1', B_2', ..., B_N'$ be the balances of the tokens in the pool after the deposit. Note that $B_j' = B_j(1+q)$ for all $j \in \{1, 2, ..., N\}$. From Equation 4.4, we know that $D_0$ and $D_1$ satisfy

$$\frac{D_0^{N+1}}{N^N \prod\limits_{j=1}^{N} B_j} + D_0 \left( AN^N - 1 \right) - AN^N \sum_{j=1}^{N} B_j = 0$$

and

$$\frac{D_1^{N+1}}{N^N \prod\limits_{j=1}^{N} B_j'} + D_1 \left( AN^N - 1 \right) - AN^N \sum_{j=1}^{N} B_j' = 0.$$

Hence,

$$0 = \frac{D_1^{N+1}}{N^N \prod\limits_{j=1}^{N} B_j'} + D_1 \left( AN^N - 1 \right) - AN^N \sum_{j=1}^{N} B_j'$$

$$= \frac{D_1^{N+1}}{N^N \prod_{j=1}^{N}\big((1+q)B_j\big)} + D_1\big(AN^N - 1\big) - AN^N \sum_{j=1}^{N}\big((1+q)B_j\big)$$

$$= (1+q)\left( \frac{\left(\dfrac{D_1}{1+q}\right)^{N+1}}{N^N \prod_{j=1}^{N} B_j} + \frac{D_1}{1+q}\big(AN^N - 1\big) - AN^N \sum_{j=1}^{N} B_j \right).$$

Let $g : \mathbb{R} \to \mathbb{R}$ be defined by

$$g(x) = \frac{x^{N+1}}{N^N \prod_{j=1}^{N} B_j} + \big(AN^N - 1\big)x - AN^N \sum_{j=1}^{N} B_j.$$

Note that both $D_0$ and $\dfrac{D_1}{1+q}$ are positive solutions of the equation $g(x) = 0$, and hence, $\dfrac{D_1}{1+q} = D_0$ since the map $g$ has exactly one positive root (see subsection 4.2.1). Thus, $D_1 = (1+q)D_0$.

We summarize the previous result in the following table:

|  | Before deposit | After deposit |
|---|---|---|
| Balance of token $j$ | $B_j$ | $(1+q)B_j$ |
| Value of parameter $D$ | $D_0$ | $(1+q)D_0$ |

**Example 4.2.** Consider a StableSwap pool with tokens USDT, USDC, and DAI. Suppose that the balances of each of these tokens are given by the following table:

|  | USDT | USDC | DAI |
|---|---|---|---|
| $B_j$ | 100,000 | 101,000 | 95,000 |

Suppose, in addition, that the current number of StableSwap LP tokens in circulation is 10,000.

If a liquidity provider wants to perform an all-asset deposit of 1% of the pool size, they will need to deposit the following amounts of each of the tokens of the pool:

$$USDT: \quad 100,000 \cdot 0.01 = 1,000$$
$$USDC: \quad 100,100 \cdot 0.01 = 1,010$$
$$DAI: \quad \ \ 95,000 \cdot 0.01 = 950.$$

In addition, they will receive an amount of 10,000.0.01 = 100 StableSwap LP tokens.

## 4.5 All-Asset Withdrawal

We will now analyze the case in which a liquidity provider redeems StableSwap LP tokens. Suppose that the liquidity provider owns an amount $m$ of StableSwap LP tokens, and let $M$ be the amount of StableSwap LP tokens in circulation. Let $\sigma = \dfrac{m}{M}$; that is, $\sigma$ is the share of the pool that the liquidity provider owns. Let $B_1$, $B_2$, ..., $B_N$ be the balances of the tokens in the pool. Clearly, in return for the amount $m$ of StableSwap LP tokens, the liquidity provider will receive, for each $j \in \{1, 2, ..., N\}$, an amount $\sigma B_j$ of token $j$.

Note also that for all $j \in \{1, 2, ..., N\}$, the balance of token $j$ in the pool after the all-asset withdrawal is $B_j - \sigma B_j = (1 - \sigma)B_j$. As we did in the case of the all-asset deposit, we will prove that the parameter $D$ changes in the same proportion as the balances of the tokens of the pool. That is, we will prove that if the value of the parameter $D$ is $D_0$ before the all-asset withdrawal, then its value after the withdrawal is $(1 - \sigma)D_0$.

We proceed in a similar way as we did in the previous section. Let $D_1$ be the value of the parameter $D$ after the withdrawal. From Equation 4.4, we know that $D_0$ and $D_1$ satisfy

$$\frac{D_0^{N+1}}{N^N \prod_{j=1}^{N} B_j} + D_0 \left( AN^N - 1 \right) - AN^N \sum_{j=1}^{N} B_j = 0$$

and

$$\frac{D_1^{N+1}}{N^N \prod_{j=1}^{N} \left( (1-\sigma) B_j \right)} + D_1 \left( AN^N - 1 \right) - AN^N \sum_{j=1}^{N} \left( (1-\sigma) B_j \right) = 0$$

Hence,

$$0 = \frac{D_1^{N+1}}{N^N \prod_{j=1}^{N} \left( (1-\sigma) B_j \right)} + D_1 \left( AN^N - 1 \right) - AN^N \sum_{j=1}^{N} \left( (1-\sigma) B_j \right)$$

$$= (1-\sigma) \left( \frac{\left( \dfrac{D_1}{1-\sigma} \right)^{N+1}}{N^N \prod_{j=1}^{N} B_j} + \frac{D_1}{1-\sigma} \left( AN^N - 1 \right) - AN^N \sum_{j=1}^{N} B_j \right)$$

Let $g : \mathbb{R} \to \mathbb{R}$ be defined by

$$g(x) = \frac{x^{N+1}}{N^N \prod_{j=1}^{N} B_j} + \left( AN^N - 1 \right) x - AN^N \sum_{j=1}^{N} B_j.$$

Note that both $D_0$ and $\dfrac{D_1}{1-\sigma}$ are positive solutions of the equation $g(x) = 0$, and hence, $\dfrac{D_1}{1-\sigma} = D_0$ since the function $g$ has exactly one positive root (see subsection 4.2.1). Thus, $D_1 = (1 - \sigma)D_0$.

We summarize the previous observations in the following table:

|  | Before withdrawal | After withdrawal |
|---|---|---|
| Balance of token $j$ | $B_j$ | $(1 - \sigma)B_j$ |
| Value of parameter $D$ | $D_0$ | $(1 - \sigma)D_0$ |

Observe that the parameter $D$ plays the same role as the parameter $V$ in Balancer (and as the parameter $L$ in Uniswap v2), meaning that for liquidity deposits (or withdrawals), the amount of tokens minted (or burned) is reflected in the change of the parameter $D$.

**Example 4.3.** Consider a StableSwap pool with tokens USDT, USDC, and DAI, and suppose that the corresponding balances are given by the following table:

|  | USDT | USDC | DAI |
|---|---|---|---|
| $B_j$ | 50,000 | 50,100 | 49,500 |

Assume that the number of StableSwap LP tokens in circulation is 100,000 and that a liquidity provider owns 100 StableSwap LP tokens. Hence, the corresponding share is $\sigma = \dfrac{100}{100,000} = 0.001$; that is, the liquidity provider owns 0.1% of the pool. If they decide to redeem their StableSwap LP tokens by performing an all-asset withdrawal, they will obtain the following amounts of each token:

$$\text{USDT}: \ 50,000 \cdot 0.001 = 50,$$
$$\text{USDC}: \ 50,100 \cdot 0.001 = 50.1,$$
$$\text{DAI}: \ 49,500 \cdot 0.001 = 49.5.$$

# 4.6 Single-Asset Withdrawal

We will now tackle the single-asset withdrawal feature of Curve Finance. Suppose that we have a StableSwap pool that has $N$ tokens with balances $B_1, B_2, ..., B_N$ and that the pool has a nonzero fee $\phi$. Assume that a liquidity provider owns a share $\sigma$ of all the StableSwap LP tokens and that instead of redeeming a proportion $\sigma$ of each of the tokens of the pool, they want to obtain all their share in terms of just one token of the pool, say, token $o$. Let $D_0$ be the value of the parameter $D$ before the withdrawal. If we were to perform an all-asset withdrawal, the value of the parameter $D$ after the withdrawal would be $(1 - \sigma)D_0$, as we saw in the previous section. Thus, it makes sense to use this value of $D$ to calculate the amount of token $o$ that will be left after the single-asset withdrawal since the parameter $D$ also reflects the liquidity of the pool in a certain sense.

Let $y$ be the amount of token $o$ after the withdrawal taking into account only the previous considerations. Since for each $j \in \{1, 2, ..., N\} - \{o\}$ the amount of token $j$ in the pool does not change after the single-asset withdrawal, applying the invariant Equation 4.2, we obtain that

$$\frac{\left((1-\sigma)D_0\right)^{N+1}}{N^N y \prod_{\substack{j=1 \\ j \neq o}}^{N} B_j} + (1-\sigma)D_0\left(AN^N - 1\right) - AN^N\left(y + \sum_{\substack{j=1 \\ j \neq o}}^{N} B_j\right) = 0$$

Proceeding as in Section 4.2, we obtain that

$$y^2 + \left(\sum_{\substack{j=1 \\ j \neq o}}^{N} B_j + \frac{(1-\sigma)D_0}{AN^N} - (1-\sigma)D_0\right)y - \frac{\left((1-\sigma)D_0\right)^{N+1}}{AN^{2N}\prod_{\substack{j=1 \\ j \neq o}}^{N} B_j} = 0.$$

Hence, we can obtain the value of $y$ by solving the previous quadratic equation. Applying Definition 4.1, observe that

$$y = \mathcal{Y}\left(o, B_1, \ldots, B_{o-1}, B_{o+1}, \ldots, B_N, (1-\sigma)D_0\right)$$

We mention that it is not important to write the explicit value of $y$ in terms of the other variables. We just want to emphasize that the value of $y$ can be computed if the values of $B_1$, $B_2$, ..., $B_N$ and the parameter $D$ are known. We summarize both pool states in the following table:

|  | Before withdrawal | After withdrawal |
|---|---|---|
| Balance of token $o$ | $B_o$ | $y$ |
| Balance of token $j$, with $j \neq o$ | $B_j$ | $B_j$ |
| Value of parameter $D$ | $D_0$ | $(1 - \sigma)D_0$ |

Therefore, if there were no fees, the liquidity provider would have to receive an amount $B_o - y$ of token o (note that $y < B_o$ by Proposition 4.1). However, since the liquidity provider receives only one token, the proportions of the tokens in the pool are modified, and thus, several spot prices also change, as occurred with the Balancer AMM. Thus, a fee has to be charged, in a similar way as when a trade is performed.

In order to charge a fair fee, we will compare the amounts of tokens in the pool after an all-asset withdrawal (AAW) and after the situation of a single-asset withdrawal (SAW) considered before.

|  | After AAW | After SAW |
|---|---|---|
| Balance of token $o$ | $(1 - \sigma)B_o$ | $y$ |
| Balance of token $j$, with $j \neq o$ | $(1 - \sigma)B_j$ | $B_j$ |
| Value of parameter $D$ | $(1 - \sigma)D_0$ | $(1 - \sigma)D_0$ |

It will be reasonable to charge the fee on the difference between the balances of the tokens in both situations, since a direct single-asset withdrawal should be equivalent to an all-asset withdrawal followed by the corresponding trades, that is, trading, for each $j \in \{1, 2, ..., N\} - \{o\}$, the amount $\sigma B_j$ of token $j$ received in the all-asset withdrawal so as to obtain token o. Hence, for each $j \in \{1, 2, ..., N\} - \{o\}$, the fee on token $j$ should be $\phi\sigma B_j$ (note that here we are considering charging fees on the way in). Therefore, if we consider the situation of an all-asset withdrawal followed by the corresponding trades, for each $j \in \{1, 2, ..., N\} - \{o\}$, the balance of token $j$ in the pool after the trades will be $(1 - \sigma)B_j + \sigma B_j = B_j$ as expected. However, in order to obtain the amount of token o that the liquidity provider should receive, we have to compute the balance of token o with respect to the balances of all the other tokens considering that the liquidity provider deposits an amount $\sigma B_j - \phi\sigma B_j$ of token $j$, due to the fees that are charged. In addition, the value of the parameter $D$ to be considered is $(1 - \sigma)D_0$, that is, the same value as after the all-asset withdrawal, since in a trade, the value of $D$ is preserved in order to compute the amount of tokens that should be given to the trader (as if there were no fees) and then when the fees are charged, the value of parameter $D$ is updated. Observe that a similar situation occurs in Uniswap v2 and in Balancer's AMMs.

This said, we have to consider the following state of the pool:

| | State |
|---|---|
| Balance of token $j$, with $j \neq o$ | $B_j - \phi\sigma B_j$ |
| Balance of token o | $y'$ |
| Value of parameter $D$ | $(1 - \sigma)D_0$ |

where the amount $y'$ has to be computed from the values of the balances of tokens $j$, with $j \neq o$ and the value of the parameter $D$ by means of the quadratic Equation 4.3. Equivalently,

$$y' = \mathcal{Y}\left(o, B_1', ..., B_{o-1}', B_{o+1}', ..., B_N', (1-\sigma)D_0\right)$$

where for each $j \in \{1, 2, \ldots, N\} - \{o\}, B'_j = B_j - \phi\sigma B_j$.

In consequence, the amount of token $o$ that has to be given to the liquidity provider should be $B_o - y'$. However, the StableSwap AMM also charges a fee on token $o$ corresponding to the difference of balances of token $o$ shown in the second table of this section. This fee is $\phi((1 - \sigma)B_o - y)$ (note that $y < (1 - \sigma)B_o$ by Proposition 4.1). In conclusion, the amount of token $o$ that is given to the liquidity provider is

$$B_o - y' - \phi\big((1-\sigma)B_o - y\big). \tag{4.6}$$

This formula can be seen in the code of the StableSwap AMM that is shown in Listing 4-4, with the following translation between the variables of the code and our notation:

$$i = o,$$
$$\texttt{D0} = D_0,$$
$$\texttt{D1} = (1 - \sigma)D_0,$$
$$\texttt{new\_y} = y,$$
$$\texttt{xp}[\texttt{j}] = B_j.$$

Note also that $y' = \texttt{self.get\_y\_D(amp, i, xp\_reduced, D1)}$.

**Listing 4-4.** Function used for the single-asset withdrawal in the smart contract of StableSwap

```
def _calc_withdraw_one_coin(_token_amount: uint256, i:
↪   int128) -> (uint256, uint256):
    # First, need to calculate
    # * Get current D
    # * Solve Eqn against y_i for D - _token_amount
    amp: uint256 = self._A()
    _fee: uint256 = self.fee * N_COINS / (4 * (N_COINS
↪   - 1))
```

```
precisions: uint256[N_COINS] = PRECISION_MUL
total_supply: uint256 = self.token.totalSupply()
xp: uint256[N_COINS] = self._xp()

D0: uint256 = self.get_D(xp, amp)
D1: uint256 = D0 - _token_amount * D0 /
↪   total_supply
xp_reduced: uint256[N_COINS] = xp

new_y: uint256 = self.get_y_D(amp, i, xp, D1)
dy_0: uint256 = (xp[i] - new_y) / precisions[i] #
↪   w/o fees

for j in range(N_COINS):
    dx_expected: uint256 = 0
    if j == i:
        dx_expected = xp[j] * D1 / D0 - new_y
    else:
        dx_expected = xp[j] - xp[j] * D1 / D0
    xp_reduced[j] -= _fee * dx_expected /
    ↪   FEE_DENOMINATOR

dy: uint256 = xp_reduced[i] - self.get_y_D(amp, i,
↪   xp_reduced, D1)
dy = (dy - 1) / precisions[i] # Withdraw less to
↪   account for rounding errors

return dy, dy_0 - dy
```

In the following example, we will compare the following two possibilities for a liquidity provider who wants to withdraw their position and obtain only one token.

- Performing a single-asset withdrawal

- Performing an all-asset withdrawal followed by some trades to obtain the token they want

**Example 4.4.** Consider a StableSwap pool having three tokens, USDT, USDC, and DAI, with the following balances:

|  | Token 1 | Token 2 | Token 3 |
|---|---|---|---|
|  | USDC | DAI | USDT |
| $B_j$ | 1,800,000 | 1,400,000 | 800,000 |

Suppose that the fee of the pool is 0.03%, the administrative pool fee is 50%, and the value of the amplification coefficient $A$ is 500.

We now apply the algorithm given by Equation 4.5 to compute the value of the parameter $D$ at the state of the pool given by the previous table, obtaining $D \approx 3,999,947.93$. This value will be called $D_0$.

Assume that a liquidity provider holds a share of 1% of the pool and that they want to withdraw their position solely in USDC. We will consider the two possibilities described previously.

- Suppose first that the liquidity provider decides to perform a single-asset withdrawal. By Equation 4.6, the amount of USDC that the liquidity provider receives is $B_o - y' - \phi((1 - \sigma)B_o - y)$, where $B_o$ is the balance of USDC in the pool, $\sigma = 0.01$, and where $y$ and $y'$ are the amounts of USDC that correspond to the following states of the pool:

|  | State A | State B |
|---|---|---|
| Balance of USDC | $y$ | $y'$ |
| Balance of DAI | 1,400,000 | $(1 - 0.0003 \cdot 0.01) \cdot 1,400,000 = 1,399,995.8$ |
| Balance of USDT | 800,000 | $(1 - 0.0003 \cdot 0.01) \cdot 800,000 = 799,997.6$ |
| Value of parameter | $0.99D_0$ | $0.99D_0$ |

or in other words,

$$y = \mathcal{Y}(1,1400000,800000,0.99D_0)$$

and

$$y' = \mathcal{Y}(1,1399995.8,799997.6,0.99D_0).$$

Solving the quadratic Equation 4.3, we obtain that $y \approx 1{,}759{,}997.36$ and $y' \approx 1{,}760{,}003.96$. Thus,

$$B_o - y' - \phi\big((1-\sigma)B_o - y\big)$$

$$\approx 1{,}800{,}000 - 1{,}760{,}003.96 - 0.0003\big(0.99 \cdot 1{,}800{,}000 - 1{,}759{,}997.36\big)$$

$$\approx 39{,}989.44$$

Hence, the liquidity provider will receive approximately 39,989.44 USDC if they decide to perform a single-asset withdrawal in USDC of their position.

- Suppose now that the liquidity provider decides to perform an all-asset withdrawal and then trade the received amounts of DAI and USDT to obtain more USDC. Clearly, they receive 18,000 USDC, 14,000 DAI, and 8,000 USDT from the all-asset withdrawal, and after it, the updated balances of the pool are as follows:

| USDC | DAI | USDT |
|---|---|---|
| 1,782,000 | 1,386,000 | 792,000 |

and the new value of the parameter $D$ is $0.99D_0 \approx 3{,}959{,}948.45$.

Now, the liquidity provider trades 14,000 DAI for USDC. The amount $A_o^{(1)}$ of USDC that they receive is

$$A_o^{(1)} = (B_o - y) \cdot (1 - \phi),$$

where the value of $y$ is obtained from Equation 4.3, or more precisely,

$$y = \mathcal{Y}(1, 1400000, 792000, 0.99 D_0) \approx 1,767,999.25.$$

Hence,

$$A_o^{(1)} = (B_o - y) \cdot (1 - \phi) \approx (1,782,000 - 1,767,999.25) \cdot 0.9997$$

$$\approx 13,996.54,$$

that is, the liquidity provider receives 13,996.54 USDC more. On the other hand, the updated balance of USDC in the pool is given by

$$B_0' = B_o - A_o^{(1)} - \omega(B_o - y)\phi$$

$$\approx 1,782,000 - 13,996.54 - 0.5 \cdot (1,782,000 - 1,767,999.25) \cdot 0.0003$$

$$\approx 1,768,001.36,$$

and the updated balances of DAI and USDT are 1,400,000 and 792,000, respectively. Also, the new value of the parameter $D$, which is obtained by applying the algorithm given by Equation 4.5, is $D_1 \approx 3,959,950.55$.

Finally, the liquidity provider trades 8,000 USDT for USDC. The amount $A_o^{(2)}$ of USDC that they receive is

$$A_o^{(2)} = (B_0' - y) \cdot (1 - \phi),$$

where $y = \mathcal{Y}\left(1,1400000,800000,D_1\right) \approx 1{,}759{,}999.46$. Thus,

$$A_o^{(2)} = \left(B_0' - y\right)\cdot\left(1 - \phi\right)$$

$$\approx \left(1{,}768{,}001.36 - 1{,}759{,}999.46\right)\cdot 0.9997$$

$$\approx 7{,}999.49$$

that is, the liquidity provider receives 7,999.49 USDC more.

Therefore, the liquidity provider has received a total amount of approximately

$$18{,}000 + 13{,}996.54 + 7{,}999.49 = 39{,}996.03 \ \ \text{USDC}$$

which is a bit more than the 39,989.44 USDC they obtain performing a single-asset withdrawal.

## 4.7  StableSwap LP Token Swap

As we did in the previous chapters, consider two different (Uniswap, Balancer, or StableSwap) pools, $P_1$ and $P_2$, respectively. Let $LP_1$ and $LP_2$ be the LP tokens of pools $P_1$ and $P_2$, respectively. We want to compute the fair swap price between $LP_1$ and $LP_2$. To this end, we will calculate the total value of each of the pools in terms of a chosen token $Z$, and then we will compute the value of tokens $LP_1$ and $LP_2$ in terms of token $Z$ so as to find the fair swap price.

Firstly, we need to make the following observation regarding StableSwap liquidity pools. Let $P$ be a StableSwap liquidity pool that has $N$ tokens $Y_1$, $Y_2$, ..., $Y_N$ with balances $B_1$, $B_2$, ..., $B_N$, respectively. Let $Z$ be any token, and for each $j \in \{1, 2, ..., N\}$, let $P_j$ be the price of token $Y_j$ in terms of token $Z$.[6] Then, the value of the whole pool $P$ in terms of token $Z$ is

$$\sum_{j=1}^{N} B_j P_j.$$

Hence, if the circulating supply of StableSwap LP tokens of pool $P$ is $M$, the value of each token is

$$\frac{1}{M} \sum_{j=1}^{N} B_j P_j.$$

Now, for $j \in \{1, 2\}$, let $V_j(Z)$ be the value of each LP token of pool $P_j$ in terms of token $Z$, which can be computed as in the previous paragraph for StableSwap pools and as in the previous chapters for Uniswap v2 or Balancer pools. Let $P$ be the price of one $LP_1$ token in terms of $LP_2$ tokens. Hence, $V_1(Z) = P \cdot V_2(Z)$, and thus,

$$P = \frac{V_1(Z)}{V_2(Z)}.$$

We will now give two examples where we will compute the fair swap price between StableSwap LP tokens and LP tokens of different pools.

---

[6] These exchange rates can be spot prices taken from DEXs or real-time prices taken from oracles.

**Example 4.5.** Suppose that Alice is a liquidity provider of a Balancer pool and Bob is a liquidity provider of a StableSwap pool. Assume that Alice's pool and Bob's pool are described by the data given in the following table:

|  | Alice | Bob |
|---|---|---|
| Pool | Balancer | StableSwap |
| Supply of pool tokens | 10,000 | 20,000 |
| Token X | ETH | USDC |
| Token Y | BTC | USDT |
| Balance of token X | 1,300 | 1,000,000 |
| Balance of token Y | 25 | 1,000,500 |
| Weight of token X | 0.8 | — |
| Weight of token Y | 0.2 | — |

We want to compute the fair swap price between Alice's and Bob's LP tokens. To this end, we will use USDC as the token $Z$ of the previous argument. Suppose that the market prices of ETH and USDT in terms of USDC are 4,000 USDC/ETH and 0.9995 USDC/USDT. Then, the value of each Balancer BPT token is

$$\frac{1}{10,000} \cdot \frac{1,300}{0.8} \cdot 4,000 = 650 \text{ USDC}$$

and the value of each StableSwap LP token is

$$\frac{1}{20,000} \cdot \left(1,000,000 + 1,000,500 \cdot 0.9995\right) \approx 100 \text{ USDC}.$$

Thus, the price of each of Alice's tokens in terms of Bob's tokens is $\frac{650}{100} = 6.5$, which means that if Alice gives Bob 1 of her tokens from the Balancer pool, she will get 6.5 of Bob's tokens from the StableSwap pool in return.

**Example 4.6.** Suppose that Alice is a liquidity provider of a Uniswap v2 pool and Bob is a liquidity provider of a StableSwap pool. Assume that Alice's pool and Bob's pool are described by the data given in the following table:

|                         | Alice       | Bob         |
| ----------------------- | ----------- | ----------- |
| Pool                    | Uniswap v2  | StableSwap  |
| Supply of pool tokens   | 20,000      | 10,000      |
| Token X                 | ETH         | USDC        |
| Token Y                 | BTC         | USDT        |
| Balance of token X      | 1,300       | 1,000,000   |
| Balance of token Y      | 100         | 1,000,500   |

Again, we want to compute the fair swap price between Alice's and Bob's tokens. As in the previous example, we will use USDC as token Z, and we will assume that the market prices of ETH and USDT in terms of USDC are 4,000 USDC/ETH and 0.9995 USDC/USDT. Then, the value of each Uniswap v2 LP token is

$$\frac{1}{20,000} \cdot (2 \cdot 1,300) \cdot 4,000 = 520 \text{ USDC}$$

and the value of each StableSwap LP token is

$$\frac{1}{10,000} \cdot (1,000,000 + 1,000,500 \cdot 0.9995) \approx 200 \text{ USDC}.$$

Thus, the price of each of Alice's tokens in terms of Bob's tokens is $\frac{520}{200} = 2.6$; that is, if Alice gives Bob 1 of her Uniswap LP tokens, she will get 2.6 of Bob's StableSwap LP tokens in return.

# 4.8 Summary

In this chapter, we explained how the invariant formula for the StableSwap AMM is obtained and how trades are performed. We also showed how liquidity can be deposited and withdrawn and gave a detailed description of the single-asset withdrawal feature.

In the next chapter, we will study the Uniswap v3 AMM, which introduces concentrated liquidity. As we shall see, the design of the Uniswap v3 AMM is very different from those of the AMMs that we have described before.

# CHAPTER 5

# Uniswap v3

Uniswap v3 turns out to be very different from Uniswap v2. Although Uniswap v3 pools also have two tokens and use a constant product formula, Uniswap v3 introduces a new concept that is called *concentrated liquidity*. The main idea of this feature is that liquidity providers can provide liquidity in a chosen price range and implies that the reserves of each position are just enough to support trading within its range. When the price goes out of that range, the position of the liquidity provider is swapped entirely into one of the two tokens, depending on the price going above or below the range.

A consequence of concentrated liquidity is that the positions are highly personalized, since liquidity providers can choose not only the amount to deposit but also the range in which they want to provide liquidity. This implies that the positions in a Uniswap v3 pool are naturally nonfungible. Therefore, liquidity providers must be given nonfungible tokens in exchange for their deposit, and these nonfungible LP tokens will have to keep a record of the details of their particular position. In addition, due to the customizable liquidity provision feature, fees must now be collected and stored separately as individual tokens rather than being automatically reinvested as liquidity into the pool.

In this chapter, we will present a thorough exposition of the Uniswap v3 AMM, together with a comprehensive analysis of liquidity provisioning in this protocol. We will also explain in detail how this AMM is actually implemented.

M. Ottina et al., *Automated Market Makers*, https://doi.org/10.1007/978-1-4842-8616-6_5

# 5.1 Ticks

As we mentioned before, in Uniswap v3, liquidity providers provide liquidity in a chosen bounded price range. However, the lower and upper limits of this range cannot be defined arbitrarily but rather can be chosen from a finite (but very big!) subset of the set of real positive numbers. The elements of this subset are called *ticks* and are indexed by integer numbers in the following way: $i \in \mathbb{Z}$ represents the tick (and hence the price) $p(i) = 1.0001^i$. We will also say that the *tick index* for the tick $p(i)$ is $i$.

Note that each tick is 0.01% away from the following tick. Uniswap v3 uses 24-bit signed integers for tick indexes. Hence, the minimum and maximum prices that it can deal with are

$$p\left(-2^{23}\right) \approx 5.07 \cdot 10^{-365}$$

and

$$p\left(2^{23} - 1\right) \approx 1.97 \cdot 10^{364}$$

which cover almost all possible prices in the asset space.

As we mentioned before, Uniswap v3 is based on the constant product formula of Uniswap v2. Recall that in a Uniswap v2 pool, the balances $A$ and $B$ of the two tokens, $X$ and $Y$, of the pool satisfy the formula $A \cdot B = L^2$, where $L$ is the liquidity parameter of the pool. Recall also that the spot price $p$ of token $X$ in terms of token $Y$ is given by $\dfrac{B}{A}$. Hence, we can express the balances $A$ and $B$ in terms of $L$ and $\sqrt{p}$ in the following way:

$$A = \frac{L}{\sqrt{p}} \quad \text{and} \quad B = L \cdot \sqrt{p}.$$

Therefore, $L$ and $\sqrt{p}$ can be used to track the state of the pool, and this is what the Uniswap v3 AMM does. Note that

$$\sqrt{p}(i)=1.0001^{\frac{i}{2}}$$

Given any price $p$, the tick index associated to $p$ is defined as the tick index of the greatest tick $t$ that satisfies $t \leq p$. Hence, the tick index $i$ associated to the price $p$ is given by

$$i=\left\lfloor \log_{1.0001} p \right\rfloor=\left\lfloor 2\log_{1.0001} \sqrt{p} \right\rfloor, \tag{5.1}$$

where for any real number $x$, $\lfloor x \rfloor$ denotes the floor of $x$, that is, the greatest integer that is less than or equal to $x$.

**Example 5.1.** Consider a Uniswap v3 pool with tokens ETH and DAI with the following balances:

| Tokens | DAI | ETH |
| --- | --- | --- |
| Balances | 40,000 | 10 |

Note that the current spot price is 4,000 DAI/ETH. This price is associated, via Equation 5.1, to the tick index

$$i=\left\lfloor \log_{1.0001} 4,000 \right\rfloor=82,944.$$

Observe that the tick index 82,944 corresponds to the price

$$1.0001^{82944} \approx 3,999.742678$$

while the tick index 82,945 corresponds to the price

$$1.0001^{82945} \approx 4,000.142653.$$

If a liquidity provider wants to provide liquidity in the price range [3700, 4300], then we can find the tick indexes that are near the limits of this interval by applying Equation 5.1:

$$\lfloor \log_{1.0001} 3{,}700 \rfloor = 82{,}164, \lfloor \log_{1.0001} 4{,}300 \rfloor = 83{,}667.$$

Now, we compute the ticks that correspond to these tick indexes and to the following ones, since the prices that we are interested in will be between those ticks. We exhibit the computations in the following table:

| Tick index | 82,164 | 82,165 | 83,667 | 83,668 |
|---|---|---|---|---|
| Tick (approx.) | 3,699.634 | 3,700.004 | 4,299.619 | 4,300.049 |

Hence, the liquidity provider can choose to provide liquidity in the price interval [3,700.004, 4,300.049].

## 5.1.1 Initialized Ticks

In Uniswap v3, ticks can have one of the following two states: initialized or uninitialized. The *initialized ticks* are those ticks that define the boundary of any of the current positions (see Figure 5-1). In other words, the initialized ticks are those ticks that we need to take care of, because the liquidity parameter can change when the price crosses the tick since a different set of positions needs to be considered. By introducing initialized ticks, the Uniswap v3 protocol avoids having to make computations and update variable values every time the price crosses a tick, since this is needed only when the price crosses an initialized tick.

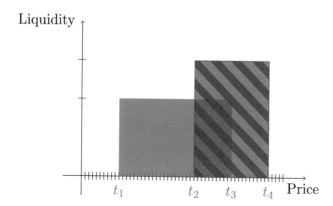

In this situation, ticks $t_1$, $t_2$, $t_3$ and $t_4$ are initialized, while all the other ticks are uninitialized. Observe that there are no initialized ticks between $t_1$ and $t_2$, or between $t_2$ and $t_3$, or between $t_3$ and $t_4$.

**Figure 5-1.** *Initialized ticks*

If a tick is used in a new position and that tick has not been initialized yet, then the tick is initialized. If a tick is a boundary of only one position, when that position is removed, the tick is uninitialized. In order to keep track of initialized and uninitialized ticks, Uniswap v3 introduces a tick bitmap, where each bitmap position corresponds to a tick index. If the tick is initialized, the value of the bitmap position that corresponds to that tick is 1, and if the tick is uninitialized, the value of the bitmap position that corresponds to that tick is 0.

## 5.1.2  Tick Spacing

In Example 5.1, we did not impose any restrictions on the tick indexes, and hence, any 24-bit signed integer could be a tick index. In general, Uniswap v3 does not allow an arbitrary choice of the tick indexes but rather introduces the concept of *tick spacing*, which, in informal terms, is a measure of the separation between the allowed tick indexes. Concretely, only tick indexes that are multiples of the tick spacing are permitted. For example, if the tick spacing is 5, the only tick indexes that can be used are the multiples of 5: ..., –10, –5, 0, 5, 10,....

In the previous example, if the tick spacing had been equal to 5, we would have had to consider the following table and the liquidity provider could have chotsen to provide liquidity in the price interval [3700.004, 4300.909].

| Tick index | 82,160 | 82,165 | 83,665 | 83,670 |
|---|---|---|---|---|
| Tick (approx.) | 3,698.155 | 3,700.004 | 4,298.759 | 4,300.909 |

The tick spacing parameter is defined and fixed when the pool is created. Clearly, if the tick spacing is small, then liquidity providers can choose more precise ranges, but on the other hand, a small tick spacing may cause trading to be more expensive in terms of gas fees, since every time the price crosses an initialized tick, new values for certain variables need to be set, which imposes a gas cost on the trader.

# 5.2 Liquidity Providers' Position

Consider a Uniswap v3 liquidity pool with tokens $X$ and $Y$. From now on, when we speak about the price, we will be referring to the price of token $X$ in terms of token $Y$. Suppose that a liquidity provider provides liquidity in a price range $[p_a, p_b]$. The main idea of Uniswap v3 is that the liquidity provider's position will use its assets to allow trading between the prices $p_a$ and $p_b$. When the price goes out of this range, the position's assets are not used for trading, and hence, the position's balances remain unmodified until the price enters the interval $[p_a, p_b]$ again. In addition, if the price falls below $p_a$, then the liquidity provider's position will be fully converted to token $X$, and if the price goes above $p_b$, then the liquidity provider's position will be fully converted to token $Y$. For example, in this last case, we can think that the whole amount of token $X$ of the position has been sold as the price of token $X$ increased, and thus, the position only has token $Y$ left. Note that in both cases, the liquidity provider is left with the asset that is less valuable.

Although the Uniswap v3 AMM is based on the constant product formula, some modifications need to take place in order to allow the balance of one of the tokens to become zero when the price reaches one of the boundaries of the interval. Concretely, we will apply the translation shown in Figure 5-2 to the curve defined by the constant product formula $xy = L^2$. Recall from subsection 2.1.1 that in this case, the price at a state $P$ coincides with the slope of the line that passes through the points $(0, 0)$ and $P$, and hence, in Figure 5-2, the price $p_a$ at the pool state $a$ is less than the price $p_b$ at the pool state $b$.

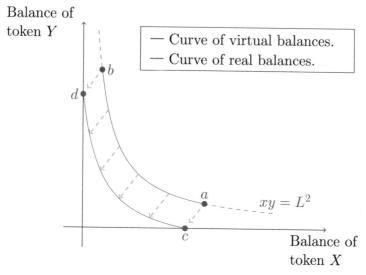

The points $a$ and $b$ represent the state of the pool defined by $xy = L^2$ when the spot price is $p_a$ and $p_b$ respectively.

***Figure 5-2.*** *Translation applied to the constant product formula in Uniswap v3*

In order to find the parameters of the translation that we need to apply, we will compute the balances of the tokens at the pool states $a$ and $b$. Let $x_a$ and $y_a$ be the balances at state $a$ of tokens $X$ and $Y$, respectively. Clearly, $x_a y_a = L^2$, and since the spot price at state $a$ is $p_a$, we obtain that $\frac{y_a}{x_a} = p_a$.

Hence, $x_a = \dfrac{L}{\sqrt{p_a}}$ and $y_a = L\sqrt{p_a}$. Thus, $a = \left( \dfrac{L}{\sqrt{p_a}}, L\sqrt{p_a} \right)$. In a similar way, we obtain that $b = \left( \dfrac{L}{\sqrt{p_b}}, L\sqrt{p_b} \right)$.

When the price is $p_a$, we need that the balance of token $Y$ is equal to zero, or equivalently, that the corresponding state is on the horizontal axis. Thus, we need to make a downward translation of an amount $y_a$. Similarly, when the price is $p_b$, we need that the balance of token $X$ is equal to zero, or equivalently, that the corresponding state is on the vertical axis. Thus, we need to make a translation to the left of an amount $x_b$. Therefore, we will make a translation of the curve defined by $xy = L^2$–which is called *curve of virtual balances*–by the vector $(-x_b, -y_a)$, that is, $\left( -\dfrac{L}{\sqrt{p_b}}, -L\sqrt{p_a} \right)$. Hence, we obtain the following equation:

$$\left( x + \frac{L}{\sqrt{p_b}} \right)\left( y + L\sqrt{p_a} \right) = L^2 \tag{5.2}$$

which defines the *curve of real balances* given in Figure 5-2.

Observe also that applying the translation mentioned previously to the points $a$ and $b$, we obtain the points $c$ and $d$ of Figure 5-2, and hence,

$$c = \left( \frac{L}{\sqrt{p_a}} - \frac{L}{\sqrt{p_b}}, 0 \right) \quad \text{and} \quad d = \left( 0, L\sqrt{p_b} - L\sqrt{p_a} \right).$$

**Virtual reserves.** We will now introduce the concept of virtual reserves, which will be needed later. With the previous notations, the *virtual reserves* (or *virtual balances*) of tokens $X$ and $Y$ are the positive real numbers $x_v$ and $y_v$ that satisfy the constant product formula $x_v y_v = L^2$ and the spot price formula $\dfrac{y_v}{x_v} = p$, where $p$ is the spot price of the pool at a certain moment. This means that the virtual reserves define a *virtual pool state* $(x_v, y_v)$ that is located on the curve of virtual balances of Figure 5-2. We can think that whenever the price belongs to the range $[p_a, p_b]$, the trades in the pool are carried out following the constant product formula $x_v y_v = L^2$, although the real reserves of the pool–the actual balances of the tokens–do not coincide with the virtual reserves.

It is important to observe that the real and virtual reserves of a liquidity provider's position ($x$ and $y$, and $x_v$ and $y_v$, respectively) are related by the following equations:

$$x_v = x + \frac{L}{\sqrt{p_b}},$$

$$y_v = y + L\sqrt{p_a},$$

(5.3)

and thus, the equations

$$\left(x + \frac{L}{\sqrt{p_b}}\right)\left(y + L\sqrt{p_a}\right) = L^2$$

and

$$x_v y_v = L^2$$

are equivalent.

Note also that if the spot price at a certain moment is $p$, since $\dfrac{y_v}{x_v} = p$, we obtain that

$$x_v = \frac{L}{\sqrt{p}} \quad \text{and} \quad y_v = L\sqrt{p}. \tag{5.4}$$

**Real reserves at price $p$.** It is important to be able to compute the balances of a liquidity provider's position when the price is $p$. Following the previous notations, we know that when $p \in [p_a, p_b]$, the balances $x$ and $y$ of the liquidity provider's position satisfy Equation 5.2, that is,

$$\left(x + \frac{L}{\sqrt{p_b}}\right)\left(y + L\sqrt{p_a}\right) = L^2.$$

If $p = p_a$, the balance $y$ of token $Y$ is equal to zero, and hence, isolating $x$ from the previous equation, we obtain that the balance of token $X$ is $\dfrac{L}{\sqrt{p_a}} - \dfrac{L}{\sqrt{p_b}}$. Thus, the state of the liquidity provider's position when $p = p_a$ is $\left(\dfrac{L}{\sqrt{p_a}} - \dfrac{L}{\sqrt{p_b}}, 0\right)$, which are the coordinates of the point $c$ of the curve of real balances of Figure 5-2 that lies on the horizontal axis. These coordinates could also have been obtained by translating the point $a$ by the vector $(-x_b, -y_a)$, which was the vector used for translating the curve of virtual balances of Figure 5-2 to the curve of real balances.

In a similar way, if $p = p_b$, the balance $x$ of token $X$ is equal to zero, and isolating $y$ from the previous equation, we obtain that the balance of token $Y$ is $L\sqrt{p_b} - L\sqrt{p_a}$. In addition, using Figure 5-2, a geometrical interpretation (similar to the previous one) can be made.

In general, for any $p \in [p_a, p_b]$, we know from Equation 5.4 that the virtual reserves are

$$x_v = \frac{L}{\sqrt{p}} \quad \text{and} \quad y_v = L\sqrt{p}.$$

Thus, applying Equation 5.3, we obtain that

$$x = \frac{L}{\sqrt{p}} - \frac{L}{\sqrt{p_b}} \quad \text{and} \quad y = L\sqrt{p} - L\sqrt{p_a}.$$

Note that in the particular cases $p = p_a$ and $p = p_b$, we obtain the results of the previous paragraphs again.

Finally, recall that when $p < p_a$, the balances of the liquidity provider's position are the same as when $p = p_a$, since these balances are not used for trades when the price goes outside the interval $[p_a, p_b]$. Similarly, when $p > p_b$, the balances of the liquidity provider's position are the same as when $p = p_b$. We condense the previous formulae into Table 5-1.

***Table 5-1.*** *Formulae for the real balances of a Uniswap v3 position*

| Price range | Real balance of token $X$ | Real balance of token $Y$ |
| --- | --- | --- |
| $p \leq p_a$ | $\dfrac{L}{\sqrt{p_a}} - \dfrac{L}{\sqrt{p_b}}$ | $0$ |
| $p_a \leq p \leq p_b$ | $\dfrac{L}{\sqrt{p}} - \dfrac{L}{\sqrt{p_b}}$ | $L\sqrt{p} - L\sqrt{p_a}$ |
| $p \geq p_b$ | $0$ | $L\sqrt{p_b} - L\sqrt{p_a}$ |

**Opening a position**. Suppose that a liquidity provider wants to provide liquidity in the price range $[p_a, p_b]$. Using the notations of this section, if the parameter $L$ is chosen, then the curve of virtual balances of Figure 5-2 is fixed, and thus, the curve of real balances is uniquely determined. Hence, by the previous results, given a price $p$, the real balances of the liquidity provider's position are determined. It is important to observe that since the trading fees of Uniswap v3 are stored separately, the parameter $L$ will remain constant, the curves of Figure 5-2 will not change, and the state of the position will always be a point of the curve of real balances.

In consequence, if the liquidity provider wants to start a position with those characteristics—$p_a$, $p_b$ and $L$ chosen—given the price $p$, they need to deposit the amounts of tokens $X$ and $Y$ given by Table 5-1 so that the state of their position belongs to the curve of real balances. Concretely, we need to consider the following three possible situations:

**(A)** If the current price $p$ is below the price range $[p_a, p_b]$ – that is, if $p < p_a$ – the liquidity provider will have to deposit only token $X$, and the amount of token $X$ they need to deposit is $\dfrac{L}{\sqrt{p_a}} - \dfrac{L}{\sqrt{p_b}}$.

**(B)** If the current price $p$ is within the price range $[p_a, p_b]$—that is, if $p_a \le p \le p_b$—the liquidity provider will need to deposit certain amounts of both tokens. Specifically, they will need to deposit an amount $\dfrac{L}{\sqrt{p}} - \dfrac{L}{\sqrt{p_b}}$ of token $X$ and an amount $L\sqrt{p} - L\sqrt{p_a}$ of token $Y$.

**(C)** If the current price $p$ is above the price range $[p_a, p_b]$—that is, if $p > p_b$—the liquidity provider will have to deposit only token $Y$, and the amount of token $Y$ they need to deposit is $L\sqrt{p_b} - L\sqrt{p_a}$.

In Figure 5-3, we can see the corresponding states of the position in the three cases discussed previously.

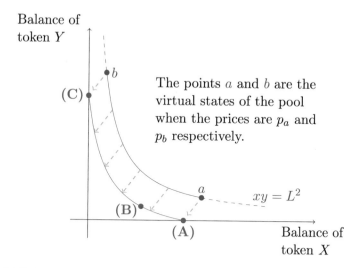

(A) If the price at the moment of the deposit is below $p_a$, the liquidity provider will have to deposit only token $X$.

(B) If the price at the moment of the deposit is between $p_a$ and $p_b$, the liquidity provider will have to deposit certain amounts of both tokens.

(C) If the price at the moment of the deposit is above $p_b$, the liquidity provider will have to deposit only token $Y$.

**Figure 5-3.** *Creating a position in Uniswap v3*

Of course, in practice, a liquidity provider will not want to choose the parameter $L$, but instead, they will choose a price range $[p_a, p_b]$ and a certain amount of either token $X$ or token $Y$ to deposit. In that case, we can simply use that amount and the formulae of Table 5-1 to find out the value of the parameter $L$, and then use this value to compute the amount of the other token that the liquidity provider needs to deposit. We show how this works in the following example.

**Example 5.2.** Suppose that a liquidity provider wants to provide liquidity in a Uniswap v3 pool with tokens ETH and USDC. Suppose, in addition, that the price of ETH in terms of USDC is 4,000 and that the liquidity provider wants to deposit liquidity in the price range (3,700.004, 4,300.049) as in Example 5.1, and that the liquidity provider wants to deposit 2 ETH.

In order to find out how much USDC the liquidity provider needs to deposit, we will first compute the value of the parameter $L$ using the formulae of the previous table. Concretely,

$$2 = \frac{L}{\sqrt{p}} - \frac{L}{\sqrt{p_b}} \approx \frac{L}{\sqrt{4,000}} - \frac{L}{\sqrt{4,300.049}} = L\left(\frac{1}{\sqrt{4,000}} - \frac{1}{\sqrt{4,300.049}}\right)$$

Hence,

$$L \approx \frac{2}{\dfrac{1}{\sqrt{4,000}} - \dfrac{1}{\sqrt{4,300.049}}} \approx 3,561.138.$$

Now we use the value of $L$ to compute the amount of USDC that is needed.

$$y = L\sqrt{p} - L\sqrt{p_a} \approx 3,561.138\left(\sqrt{4,000} - \sqrt{3,700.004}\right) \approx 8,610.458.$$

Therefore, the liquidity provider will have to make a deposit of 2 ETH and 8,610.458 USDC to set up their position.

**Value of a position.** We will now compute the value of a liquidity provider's position in terms of the price $p$, which will be denoted by $V(p)$. If $x_p$ and $y_p$ denote the real reserves of the liquidity provider's position when the price is $p$, then $V(p) = x_p p + y_p$. Note that when the position is set up, the value of the liquidity parameter $L$ and the price range $[p_a, p_b]$ in which liquidity will be provided are fixed. Using the results of Table 5-1, we obtain the following:

- If $p \le p_a$, then

$$V(p) = \left( \frac{L}{\sqrt{p_a}} - \frac{L}{\sqrt{p_b}} \right) p = L \left( \frac{1}{\sqrt{p_a}} - \frac{1}{\sqrt{p_b}} \right) p.$$

- If $p_a \le p \le p_b$, then

$$V(p) = \left( \frac{L}{\sqrt{p}} - \frac{L}{\sqrt{p_b}} \right) p + L\sqrt{p} - L\sqrt{p_a} = L \left( 2\sqrt{p} - \frac{p}{\sqrt{p_b}} - \sqrt{p_a} \right).$$

- If $p \ge p_b$, then

$$V(p) = L\sqrt{p_b} - L\sqrt{p_a} = L \left( \sqrt{p_b} - \sqrt{p_a} \right).$$

We use the previous formulae to plot, in Figure 5-4, the value of a Uniswap v3 liquidity provider's position as a function of the price $p$.

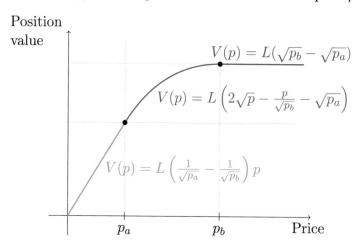

**Figure 5-4.** *Value of a liquidity provider's position as a function of the price p*

**Example 5.3.** Consider the liquidity provider's position of Example 5.2, with liquidity parameter $L \approx 3{,}561.138$ and price range $(3700.004, 4300.049)$. The liquidity provider deposited 2 ETH and approximately 8,610.458 USDC when the price of ETH in terms of USDC was 4,000, so the initial value of their position is

$$V(4{,}000) = 2 \cdot 4{,}000 + 8{,}610.458 = 16{,}610.458 \text{ USDC.}$$

If the price goes up to 4,200, from the formulae of Table 5-1, we obtain that the real balance of ETH in the liquidity provider's position will be approximately

$$\frac{3{,}561.138}{\sqrt{4{,}200}} - \frac{3{,}561.138}{\sqrt{4{,}300.049}} \approx 0.643$$

and the real balance of USDC in their position will be approximately

$$3{,}561.138\sqrt{4{,}200} - 3{,}561.138\sqrt{3{,}700.004} \approx 14{,}172.435$$

Thus,

$$V(4{,}200) \approx 0.643 \cdot 4{,}200 + 14{,}172.435 = 16{,}873.035 \text{ USDC.}$$

Observe that if the liquidity provider had held their assets instead of depositing them into the pool, their value would have been approximately

$$2 \cdot 4{,}200 + 8{,}610.458 = 17{,}010.458 \text{ USDC}$$

which is greater than $V(4{,}200)$. Thus, the liquidity provider is facing an impermanent loss. We will analyze impermanent losses in Uniswap v3 in the following section.

# 5.3 Impermanent Loss

In the previous section, we studied and computed the value of a liquidity provider's position in terms of the price $p$. We will now study the possible impermanent losses that may occur, generalizing what was shown in Example 5.3.

As in the previous section, consider a Uniswap v3 liquidity pool with tokens $X$ and $Y$ and a liquidity provider that deposits liquidity into the pool in a range $[p_a, p_b]$. Let $p_0$ be the entry price (i.e., the price at which the liquidity deposit is made); let $x_0$ and $y_0$ be the amounts of tokens $X$ and $Y$, respectively, that the liquidity provider deposits; and let $L$ be the corresponding liquidity parameter. Let $p$ be a positive real number, and let $x_p$ and $y_p$ be the real reserves of tokens $X$ and $Y$, respectively, that correspond to the liquidity provider's position when the price of token $X$ in terms of token $Y$ is $p$. Let $V(p)$ be the value of the liquidity provider's position (in terms of token $Y$) and let $W(p)$ be the current joint value of the amounts $x_0$ and $y_0$ of tokens $X$ and $Y$, respectively (in terms of token $Y$), that is the value of the liquidity provider's position if they had not deposited it into the pool. Note that $V(p) = x_p \cdot p + y_p$ and $W(p) = x_0 \cdot p + y_0$.

Recall also that the fraction of impermanent loss is computed as

$$\mathrm{IL}(p) = \frac{V(p)}{W(p)} - 1.$$

We will divide our analysis into three cases with three subcases each.

<u>Case 1</u>: $p_0 \in [p_a, p_b]$

Applying the formulae of Table 5-1, we obtain that

$$x_0 = L \cdot \left( \frac{1}{\sqrt{p_0}} - \frac{1}{\sqrt{p_b}} \right) \text{ and } y_0 = L \cdot \left( \sqrt{p_0} - \sqrt{p_a} \right).$$

- If $p \in [p_a, p_b]$, from Table 5-1, we know that

$$x_p = L \cdot \left( \frac{1}{\sqrt{p}} - \frac{1}{\sqrt{p_b}} \right) \text{ and } y_p = L \cdot \left( \sqrt{p} - \sqrt{p_a} \right).$$

Thus,

$$
\begin{aligned}
\mathrm{IL}(\mathrm{p}) \;&= \frac{V(p)}{W(p)} - 1 \\[2mm]
&= \frac{L \cdot \left( \dfrac{1}{\sqrt{p}} - \dfrac{1}{\sqrt{p_b}} \right) \cdot p + L \cdot \left( \sqrt{p} - \sqrt{p_a} \right)}{L \cdot \left( \dfrac{1}{\sqrt{p_0}} - \dfrac{1}{\sqrt{p_b}} \right) \cdot p + L \cdot \left( \sqrt{p_0} - \sqrt{p_a} \right)} - 1 \\[2mm]
&= \frac{2\sqrt{p} - \dfrac{p}{\sqrt{p_b}} - \sqrt{p_a}}{\dfrac{p}{\sqrt{p_0}} - \dfrac{p}{\sqrt{p_b}} + \sqrt{p_0} - \sqrt{p_a}} - 1.
\end{aligned}
$$

- If $p \leq p_a$, $y_p = 0$, and from Table 5-1, we have that

$$x_p = L \cdot \left( \frac{1}{\sqrt{p_a}} - \frac{1}{\sqrt{p_b}} \right).$$

Thus,

$$\text{IL}(p) \quad = \frac{V(p)}{W(p)} - 1$$

$$= \frac{L \cdot \left( \dfrac{1}{\sqrt{p_a}} - \dfrac{1}{\sqrt{p_b}} \right) \cdot p}{L \cdot \left( \dfrac{1}{\sqrt{p_0}} - \dfrac{1}{\sqrt{p_b}} \right) \cdot p + L \cdot \left( \sqrt{p_0} - \sqrt{p_a} \right)} - 1$$

$$= \frac{\dfrac{p}{\sqrt{p_a}} - \dfrac{p}{\sqrt{p_b}}}{\dfrac{p}{\sqrt{p_0}} - \dfrac{p}{\sqrt{p_b}} + \sqrt{p_0} - \sqrt{p_a}} - 1.$$

- If $p \geq p_b$, $x_p = 0$, and from Table 5-1, we know that

$$y_p = L \cdot \left( \sqrt{p_b} - \sqrt{p_a} \right).$$

Thus,

$$\text{IL}(p) \quad = \frac{V(p)}{W(p)} - 1$$

$$= \frac{L \cdot \left( \sqrt{p_b} - \sqrt{p_a} \right)}{L \cdot \left( \dfrac{1}{\sqrt{p_0}} - \dfrac{1}{\sqrt{p_b}} \right) \cdot p + L \cdot \left( \sqrt{p_0} - \sqrt{p_a} \right)} - 1$$

$$= \frac{\sqrt{p_b} - \sqrt{p_a}}{\dfrac{p}{\sqrt{p_0}} - \dfrac{p}{\sqrt{p_b}} + \sqrt{p_0} - \sqrt{p_a}} - 1.$$

In Figure 5-5, we plot the impermanent loss of a Uniswap v3 position in the case that the entry price $p_0$ belongs to the price interval $[p_a, p_b]$, and we compare it with that of a Uniswap v2 position. As we can see, the impermanent loss is much higher for Uniswap v3 positions.

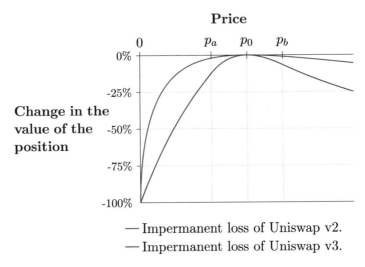

— Impermanent loss of Uniswap v2.
— Impermanent loss of Uniswap v3.

**Figure 5-5.** *Case 1: Impermanent loss of a Uniswap v3 position when the entry price belongs to the chosen interval compared with the impermanent loss of Uniswap v2*

Case 2: $p_0 \leq p_a$.
From Table 5-1, we know that

$$x_0 = L \cdot \left( \frac{1}{\sqrt{p_a}} - \frac{1}{\sqrt{p_b}} \right) \text{ and } y_0 = 0.$$

- If $p \leq p_a$, $y_p = 0$, and from Table 5-1, we know that

$$x_p = L \cdot \left( \frac{1}{\sqrt{p_a}} - \frac{1}{\sqrt{p_b}} \right).$$

Thus,

$$IL(p) = \frac{V(p)}{W(p)} - 1 = \frac{L \cdot \left(\dfrac{1}{\sqrt{p_a}} - \dfrac{1}{\sqrt{p_b}}\right) \cdot p}{L \cdot \left(\dfrac{1}{\sqrt{p_a}} - \dfrac{1}{\sqrt{p_b}}\right) \cdot p} - 1 = 1 - 1 = 0.$$

Note that there is no impermanent loss in this case since the deposited position coincides with the real reserves.

- If $p \in [p_a, p_b]$, from Table 5-1, we have that

$$x_p = L \cdot \left(\frac{1}{\sqrt{p}} - \frac{1}{\sqrt{p_b}}\right) \text{ and } y_p = L \cdot \left(\sqrt{p} - \sqrt{p_a}\right).$$

Thus,

$$IL(p) = \frac{V(p)}{W(p)} - 1$$

$$= \frac{L \cdot \left(\dfrac{1}{\sqrt{p}} - \dfrac{1}{\sqrt{p_b}}\right) \cdot p + L \cdot \left(\sqrt{p} - \sqrt{p_a}\right)}{L \cdot \left(\dfrac{1}{\sqrt{p_a}} - \dfrac{1}{\sqrt{p_b}}\right) \cdot p} - 1$$

$$= \frac{2\sqrt{p} - \dfrac{p}{\sqrt{p_b}} - \sqrt{p_a}}{\dfrac{p}{\sqrt{p_a}} - \dfrac{p}{\sqrt{p_b}}} - 1.$$

- If $p \geq p_b$, $x_p = 0$, and from Table 5-1, we know that

$$y_p = L \cdot \left(\sqrt{p_b} - \sqrt{p_a}\right).$$

Thus,

$$
\begin{aligned}
\mathrm{IL}(p) \quad &= \frac{V(p)}{W(p)} - 1 = \frac{L \cdot \left(\sqrt{p_b} - \sqrt{p_a}\right)}{L \cdot \left(\dfrac{1}{\sqrt{p_a}} - \dfrac{1}{\sqrt{p_b}}\right) \cdot p} - 1 \\[2ex]
&= \frac{\sqrt{p_b} - \sqrt{p_a}}{\dfrac{p}{\sqrt{p_a}} - \dfrac{p}{\sqrt{p_b}}} - 1 = \frac{\sqrt{p_b} - \sqrt{p_a}}{p \cdot \dfrac{\sqrt{p_b} - \sqrt{p_a}}{\sqrt{p_a}\sqrt{p_b}}} - 1 \\[2ex]
&= \frac{\sqrt{p_a}\sqrt{p_b}}{p} - 1.
\end{aligned}
$$

In a similar way as in the previous case, in Figure 5-6, we plot the impermanent loss of a Uniswap v3 position in the case that $p_0 \leq p_a$, and we compare it with that of the Uniswap v2 case. As we can see, the impermanent loss for Uniswap v3 positions is much higher when the price is above the midpoint of the price interval but is lower when the price is near $p_a$ and is zero when the price is below $p_a$.

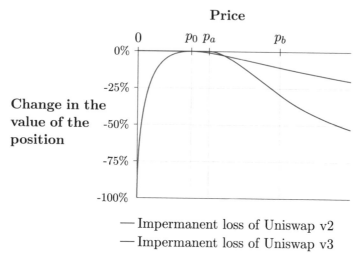

**Figure 5-6.** *Case 2. Impermanent loss of a Uniswap v3 position when the entry price is below the chosen interval compared with the impermanent loss of Uniswap v2*

<u>Case 3</u>: $p_0 \geq p_b$.

From Table 5-1, we know that

$$x_0 = 0 \text{ and } y_0 = L \cdot \left( \sqrt{p_b} - \sqrt{p_a} \right)$$

- If $p \leq p_a$, $y_p = 0$, and from Table 5-1, we know that

$$x_p = L \cdot \left( \frac{1}{\sqrt{p_a}} - \frac{1}{\sqrt{p_b}} \right)$$

Thus,

$$\mathrm{IL}(p) = \frac{V(p)}{W(p)} - 1 \;=\; \frac{L \cdot \left( \dfrac{1}{\sqrt{p_a}} - \dfrac{1}{\sqrt{p_b}} \right) \cdot p}{L \cdot \left( \sqrt{p_b} - \sqrt{p_a} \right)} - 1$$

$$= \frac{\left( \dfrac{1}{\sqrt{p_a}} - \dfrac{1}{\sqrt{p_b}} \right) \cdot p}{\sqrt{p_b} - \sqrt{p_a}} - 1 = \frac{\left( \dfrac{\sqrt{p_b} - \sqrt{p_a}}{\sqrt{p_a} \sqrt{p_b}} \right) \cdot p}{\sqrt{p_b} - \sqrt{p_a}} - 1$$

$$= \frac{p}{\sqrt{p_a} \sqrt{p_b}} - 1.$$

- If $p \in [p_a, p_b]$, from Table 5-1, we have that

$$x_p = L \cdot \left( \frac{1}{\sqrt{p}} - \frac{1}{\sqrt{p_b}} \right) \text{ and } y_p = L \cdot \left( \sqrt{p} - \sqrt{p_a} \right).$$

Thus,

$$\mathrm{IL}(p) \;= \frac{V(p)}{W(p)} - 1$$

$$= \frac{L \cdot \left( \dfrac{1}{\sqrt{p}} - \dfrac{1}{\sqrt{p_b}} \right) \cdot p + L \cdot \left( \sqrt{p} - \sqrt{p_a} \right)}{L \cdot \left( \sqrt{p_b} - \sqrt{p_a} \right)} - 1$$

$$= \frac{2\sqrt{p} - \dfrac{p}{\sqrt{p_b}} - \sqrt{p_a}}{\sqrt{p_b} - \sqrt{p_a}} - 1.$$

- If $p \geq p_b$, $x_p = 0$, and from Table 5-1, we know that

$$y_p = L \cdot \left( \sqrt{p_b} - \sqrt{p_a} \right).$$

Thus,

$$IL(p) = \frac{V(p)}{W(p)} - 1 = \frac{L \cdot \left( \sqrt{p_b} - \sqrt{p_a} \right)}{L \cdot \left( \sqrt{p_b} - \sqrt{p_a} \right)} - 1 = 1 - 1 = 0.$$

Again, note that there is no impermanent loss in this case since the deposited position coincides with the real reserves.

In a similar way as in the previous cases, we plot in Figure 5-7 the impermanent loss of a Uniswap v3 position in the case that $p_0 \geq p_b$, and we compare it with that of a Uniswap v2 one. As we can see, the impermanent loss for Uniswap v3 positions is relatively higher when the price is below the midpoint of the price interval but is lower when the price is near $p_b$ and is zero when the price is above $p_b$.

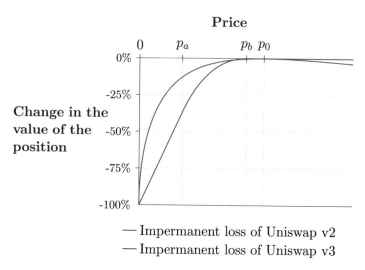

*Figure 5-7.*  *Case 3: Impermanent loss of a Uniswap v3 position when the entry price is above the chosen interval compared with the impermanent loss of Uniswap v2*

We need to mention that in this case, the comparison between Uniswap v3 and v2 is tricky when token $Y$ is a stablecoin, and thus, we will give a more detailed analysis. Consider liquidity pools with ETH and USDC and suppose that the entry price is above the chosen interval. In a Uniswap v3 pool, the liquidity provider would have to deposit only USDC, but in a Uniswap v2 pool, they would have to swap half of that USDC into ETH before making the deposit. Hence, if the price of ETH in terms of USDC goes up, the position of the Uniswap v3 pool will not change–it will have the same amount of USDC and 0 ETH–while the position of the Uniswap v2 pool will have less ETH than at the beginning and more USDC, which amounts to more than half of USDC and less than half of ETH (with respect to the initial value). Since the price of ETH increased, this gives an impermanent loss with respect to the position that has half of USDC and half of ETH, but nevertheless, this position would have more value than the Uniswap v3 position that consisted of simply holding USDC.

# 5.4 Multiple Positions

In this section, we will explain how multiple different positions are combined in Uniswap v3. Concretely, we will show that the liquidity parameter at a price $p$ that is not an initialized tick is the sum of the liquidity parameters of all the positions whose price range contains $p$. Observe that the liquidity parameter at an initialized tick cannot be defined in a reasonable way since the liquidity parameters of a Uniswap v3 pool at the right and at the left of an initialized tick might be different.

**Proposition 5.1.** *Let $n \in \mathbb{N}$. Suppose that exactly $n$ positions exist in a Uniswap v3 pool. For each $j \in \{1, 2, ..., n\}$, let $[a_j, b_j]$ be the price range of position $j$ and let $L_j$ be its liquidity parameter. Let $T = \{a_1, a_2, ..., a_n\} \cup \{b_1, b_2, ..., b_n\}$. Let $p_0$ be a positive real number such that $p_0 \notin T$ and let*

$$A = \left\{ j \in \{1, 2, ..., n\} \mid p_0 \in \left[ a_j, b_j \right] \right\}.$$

*Then the liquidity parameter of the Uniswap v3 pool at the price*
$p_0$ *is* $\sum_{j \in A} L_j$.

*Proof.* Observe that if $A = \varnothing$, then $p_0$ does not belong to any of the price ranges of the positions of the pool, and hence, the liquidity parameter of the pool at a price $p_0$ is 0, which coincides with $\sum_{j \in A} L_j$, since the last one is a sum without summands.

Therefore, we may assume that $A \neq \varnothing$. Note that if $j \notin A$, then position $j$ does not provide liquidity at the price $p_0$, and thus, we do not need to consider position $j$.

Let

$$I = \bigcap_{j \in A} \left[ a_j, b_j \right].$$

Clearly, $I$ is an interval and $p_0 \in I$. For each $j \in A$, let $x_j(p)$ and $y_j(p)$ denote the virtual reserves of position $j$ as functions of the price $p$. Note that for each $j \in A$, the equation $x_j(p) y_j(p) = L_j^2$ is valid for prices within the interval $[a_j, b_j]$ and, in particular, for prices that belong to the interval $I$ since $I \subseteq [a_j, b_j]$. Note also that by the spot price formula, we have that

$$\frac{y_j(p)}{x_j(p)} = p$$

for all $j \in A$ and for all $p \in I$. Therefore,

$$y_j(p) = p x_j(p) \quad \text{and} \quad x_j(p) = \frac{L_j}{\sqrt{p}}$$

for all $j \in A$ and for all $p \in I$.

Since trading is done using the equation of virtual reserves and since the positions indexed by $j \in A$ provide liquidity in the interval $I$, it follows that when $p \in I$, the virtual reserves that we need to consider are

$$\sum_{j \in A} x_j(p) \text{ and } \sum_{j \in A} y_j(p).$$

Let $p \in I$. We have that

$$\left( \sum_{j \in A} x_j(p) \right)\left( \sum_{j \in A} y_j(p) \right) = \left( \sum_{j \in A} x_j(p) \right)\left( \sum_{j \in A} p x_j(p) \right) =$$

$$= p\left( \sum_{j \in A} x_j(p) \right)^2 = p\left( \sum_{j \in A} \frac{L_j}{\sqrt{p}} \right)^2 =$$

$$= \left( \sum_{j \in A} L_j \right)^2.$$

Therefore, when the price $p$ belongs to the interval $I$, trading in the Uniswap v3 pool that we are considering is equivalent to trading in a constant product AMM with liquidity parameter $\sum_{j \in A} L_j$. The result follows since the liquidity parameter $\sum_{j \in A} L_j$ is valid within the interval $I$, hence, in particular, at price $p_0$ as $p_0 \in I$. □

In Figure 5-8, we illustrate a simple case of a Uniswap v3 pool with only two positions that provide liquidity in the price intervals $[p_1, p_3]$ and $[p_2, p_4]$ with liquidity parameters $L_1$ and $L_2$, respectively. Clearly, the price boundaries $p_1, p_2, p_3$ and $p_4$ are chosen among the permitted ticks. Note that in the intersection of both intervals, the liquidity parameter of the pool is $L_1 + L_2$. Observe also that in the graph of the curves of virtual balances, the gray dashed lines represent states with a certain price $p_j$ (this follows from the spot price formula).

**Liquidity provided
by each position.**

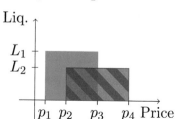

**Total liquidity.**

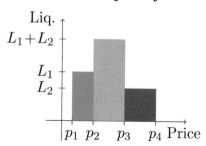

**Curves of virtual balances.**

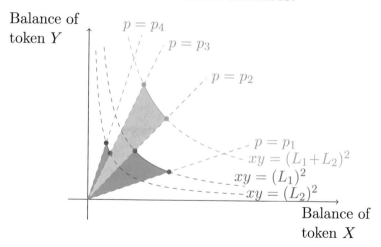

***Figure 5-8.*** *Combining two positions in Uniswap v3*

# 5.5 Protocol Implementation

In Uniswap v2, a trading fee is charged on the incoming token for each
trade that is made. The collected fees stay in the pool, causing the liquidity
parameter $L$ to increase. And liquidity providers collect their share of fees
when they remove liquidity.

In Uniswap v3, things are completely different. Fees are still charged on the incoming token, but they are stored separately and kept outside the pool as the individual tokens in which they are paid. This means that fees are not reinvested as in Uniswap v2. In addition, liquidity providers earn a portion of the trading fees only when the price moves within the range in which they provide liquidity. In order to allow this, the smart contract of Uniswap v3 needs to keep track of many variables.

In this section, we will explain how the Uniswap v3 protocol is actually implemented. To this end, we will analyze several concepts since the implementation of the Uniswap v3 protocol is much more complex than those of the AMMs that we have studied in the previous chapters.

## 5.5.1 Variables

We will begin by describing several variables that are used in the smart contract of Uniswap v3 and explaining how they are used to keep track of different things. We will be interested in the variables given in Table 5-2. The variables indicated with (T) depend on the tick index, and hence, we will have one of those variables for each initialized tick. We will always denote such variables as a function of the tick index $i$ in order to indicate such dependence.

***Table 5-2.*** *Variables used in the smart contract of Uniswap v3*

| Variable | Notation | Type |
|---|---|---|
| Current tick index | $i_c$ | |
| Square root of the current price | $\sqrt{p}$ | |
| Total liquidity | $L_{tot}$ | |
| Net liquidity | $\Delta L(i)$ | (T) |
| Gross liquidity | $L_g(i)$ | (T) |
| Total of collected fees | $f_g^0, f_g^1$ | |
| Fees collected from "outside" | $f_o^0(i), f_o^1(i)$ | (T) |

**Current tick index.** The current tick index variable $i_c$ gives the tick index that corresponds to the current price. Concretely, if $p$ is the current price, then, from Equation 5.1, we obtain that

$$i_c = \lfloor \log_{1.0001} p \rfloor.$$

**Square root of the current price** and **total liquidity.** Instead of tracking the pool balances as in Uniswap v2, the smart contract of Uniswap v3 tracks the square root of the price $\sqrt{p}$ and the total liquidity $L_{tot}$ at that specific price. As we have seen in Section 5.2, the values of the virtual reserves $x_v$ and $y_v$ can be obtained from $\sqrt{p}$ and $L_{tot}$, and hence, the required amounts of each token for any given trade can be computed from those values. Therefore, knowing the values of $\sqrt{p}$ and $L_{tot}$ is enough to compute the parameters of a trade. Moreover, tracking the values of $\sqrt{p}$ and $L_{tot}$ is equivalent to tracking the virtual reserves $x_v$ and $y_v$.

**Net liquidity.** For any initialized tick index $i$, the net liquidity variable $\Delta L(i)$ measures the change of the liquidity parameter $L_{tot}$ when crossing the initialized tick whose index is $i$. Hence, when the price crosses the tick index $i$, the amount $\Delta L(i)$ is added to or subtracted from $L_{tot}$ in order to obtain the liquidity parameter of the new price interval.

For example, suppose that the initialized ticks are $t_1$, $t_2$, $t_3$, $t_4$, and $t_5$ with $t_1 < t_2 < t_3 < t_4 < t_5$. For each $j \in \{1,2,3,4,5\}$, let $i_j$ be the tick index of $t_j$. Suppose that three liquidity providers, $A$, $B$, and $C$, provided liquidity in the intervals $[t_1, t_3]$, $[t_2, t_4]$, and $[t_4, t_5]$, respectively, with liquidity parameters $L_A$, $L_B$, and $L_C$ (see Figure 5-9). If we suppose that we are moving through the prices from left to right (i.e., from lower prices to higher prices), then, when we cross tick $t_1$, an amount $L_A$ of liquidity is added. Thus, we set $\Delta L(i_1) = L_A$. Afterward, when we cross tick $t_2$, an amount $L_B$ of liquidity is added, and hence, we define $\Delta L(i_2) = L_B$. Then, when we cross tick $t_3$, an amount $L_A$ of liquidity is removed, so we set $\Delta L(i_3) = -L_A$. After that, when we cross tick $t_4$, an amount $L_B$ of liquidity is removed, and an amount $L_C$ of liquidity is added. Hence, we define $\Delta L(i_4) = L_C - L_B$. Finally, when we cross tick $t_5$, an amount $L_C$ of liquidity is removed, and we set $\Delta L(i_5) = -L_C$.

## Liquidity provided by each position.

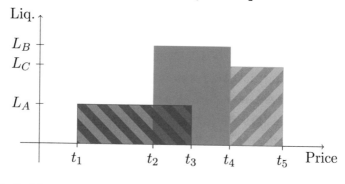

## Total liquidity.

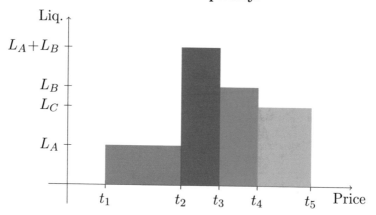

**Figure 5-9.** *Example of total liquidity in a Uniswap v3 pool with three positions*

In this way, the smart contract keeps track of the total liquidity $L_{tot}$, and when the price crosses an initialized tick $t$ whose index is $i$, the total liquidity is updated as $L_{tot} + \Delta L(i)$ if the tick $t$ is crossed from left to right, or as $L_{tot} - \Delta L(i)$ if the tick $t$ is crossed from the right to left.

**Gross liquidity.** For any initialized tick index $i$, the gross liquidity at tick $i$ is the sum of the liquidity parameters of all the positions referencing the tick indexed by $i$. This variable is used to track which ticks are really needed. If, after a position is removed, the gross liquidity of an initialized tick $t$ becomes 0, this means that tick $t$ is no longer referenced by any position, and thus, it can be uninitialized.

Observe that this cannot be done simply by observing the net liquidity of a tick, since the net liquidity of a tick $t$ might be zero, but there might still be positions referencing that tick. For example, if there are exactly two positions referencing tick $t$ given by price intervals $[p_1, t]$ and $[t, p_2]$, respectively, and both positions have the same liquidity parameter, then, if $i$ is the tick index of $t$, we obtain that the net liquidity $\Delta L(i)$ is 0, but clearly, we still need to keep track of tick $t$ because there are two positions that reference it and, for example, we need to record separately the fees earned by both positions.

Before going on with the description of the next variables, we will give an example to illustrate how the net liquidity and gross liquidity variables are defined.

**Example 5.4** (Net and gross liquidities). Consider a Uniswap v3 liquidity pool with just four positions, whose price ranges and liquidity parameters are represented in Figure 5-10. Note that the only initialized ticks are $t_1$, $t_2$, $t_3$, $t_4$, $t_5$, and $t_6$.

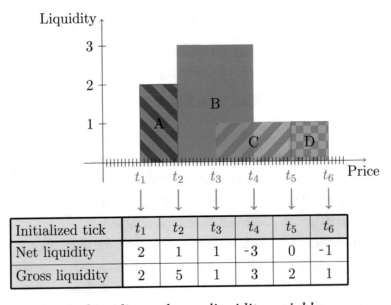

| Initialized tick | $t_1$ | $t_2$ | $t_3$ | $t_4$ | $t_5$ | $t_6$ |
|---|---|---|---|---|---|---|
| Net liquidity | 2 | 1 | 1 | -3 | 0 | -1 |
| Gross liquidity | 2 | 5 | 1 | 3 | 2 | 1 |

**Figure 5-10.** *Net liquidity and gross liquidity variables*

First, we will analyze the net liquidity variable. Moving from lower prices to higher prices, when we cross tick $t_1$, an amount 2 of liquidity is added (the liquidity of position A). Thus, the value of the net liquidity variable at tick $t_1$ is 2. Then, when we cross tick $t_2$, an amount 3 of liquidity is added (the liquidity of position B), and an amount 2 of liquidity is removed (the liquidity of position A). Hence, the value of the net liquidity variable at tick $t_2$ is 1. Note that the total liquidity within the interval $(t_2, t_3)$ is 3, which is 1 unit higher than the total liquidity that corresponds to the interval $(t_1, t_2)$.

Afterward, when we cross tick $t_3$, an amount 1 of liquidity is added (the liquidity of position C), and hence, the value of the net liquidity variable at tick $t_3$ is 1. Then, when we cross tick $t_4$, an amount 3 of liquidity is removed (the liquidity of position B), and hence, the value of the net liquidity variable at tick $t_4$ is –3. Observe that the total liquidity within the interval $(t_4, t_5)$ is 1, which is 3 units lower than the total liquidity that corresponds to the interval $(t_3, t_4)$ (which is equal to 4 since it is the sum of the liquidity parameters of positions B and C).

Then, when we cross tick $t_5$, an amount 1 of liquidity is removed (the liquidity of position C), and an amount 1 of liquidity is added (the liquidity of position D). Thus, the value of the net liquidity variable at tick $t_5$ is 0. Finally, when we cross tick $t_6$, an amount 1 of liquidity is removed, and hence, the value of the net liquidity variable at tick $t_6$ is $-1$.

Now, we will analyze the gross liquidity variable. Clearly, the value of the gross liquidity variable at tick $t_1$ is 2 since the only position that references tick $t_1$ is position A, and the liquidity parameter of position $A$ is 2. Next, the value of the gross liquidity variable at tick $t_2$ is 5 since the positions that reference tick $t_2$ are A and B, and their liquidity parameters are 2 and 3, respectively. Then, observe that the only position that references tick $t_3$ is position C, and hence, the value of the gross liquidity variable at tick $t_3$ is 1, which is the value of the liquidity parameter of position C. Note that position B does not reference tick $t_3$ even though tick $t_3$ is within its range. In a similar way, the only position that references tick

$t_4$ is position B, and hence, the value of the gross liquidity variable at tick $t_4$ is 3 since this is the value of the liquidity parameter of position B.

Next, there exist two positions that reference tick $t_5$, which are positions C and D. Since the liquidity parameters of both of them are 1, we obtain that the value of the gross liquidity variable at tick $t_5$ is 2. Observe that the value of the net liquidity variable at tick $t_5$ is 0, but nonetheless, there exist two positions that reference that tick. This illustrates why the gross liquidity variable is needed in the smart contract of Uniswap v3. Finally, the value of the gross liquidity variable at tick $t_6$ is 1 since the only position that references tick $t_6$ is position D, which has liquidity parameter equal to 1.

**Total of collected fees.** The protocol keeps track of all the fees that were collected. These fees are not deposited into the pool but rather kept separately in the same token that they were charged. Hence, two variables are needed to do that, which are denoted by $f_g^0$ and $f_g^1$. For simplicity, the variables $f_g^0$ and $f_g^1$ record the total amount of fees collected per unit of liquidity. Otherwise, it would be harder to track the amount of fees that a particular liquidity provider owns if the amount of total liquidity changes, and in this case, it would be needed to record the amount of fees owned by each and every position.

**Total of fees collected from "outside."** Any tick $t$ divides the set of positive real numbers into two intervals: $[0, t)$ and $[t, +\infty)$. Given any tick $t$ and the current price $p$, we define the outer interval that corresponds to tick $t$ with respect to the price $p$ as the only of the two intervals, $[0, t)$ and $[t, +\infty)$, that does not contain $p$. Clearly, which of the previous intervals is the outer one may change if the price changes.

For example, consider tick $t = 3{,}699.634$ with tick index $82{,}164$ from Example 5.1. The tick $t$ divides the set of positive real numbers into the two intervals $[0, 3{,}699.634)$ and $[3{,}699.634, +\infty)$. If the current price $p$ is $3{,}750$, then the outer interval that corresponds to tick $t$ with respect to $p$ is $[0, 3{,}699.634)$, since that interval does **not** contain $p$. On the other hand, if the current price $p$ is 3,345, then the outer interval that corresponds to tick $t$ with respect to $p$ is $[3,699.634, +\infty)$, since that interval does **not** contain $p$.

For any initialized tick index $i$, the smart contract of Uniswap v3 tracks two variables, denoted by $f_o^0(i)$ and $f_o^1(i)$, which record the total amount of fees per unit of liquidity that were collected in each of the two tokens in the outer interval corresponding to the tick with index $i$, that is, in the interval $[0, t)$ if $i_c \geq i$, or in the interval $[t, +\infty)$ if $i_c < i$.

Given an initialized tick $t$ with index $i$, if the price crosses $t$, then the outer interval changes, and hence, the variables $f_o^0(i)$ and $f_o^1(i)$ need to be updated consequently. This is done by setting, for each $j \in \{0, 1\}$, the variable $f_o^j(i)$ as $f_g^j - f_o^j(i)$, since the sum of the amount of fees collected in the intervals $[0, t)$ and $[t, +\infty)$ is equal to the total collected fees.

We also need to mention that when a new tick index $i$ is initialized, the values of $f_o^j(i)$ for $j \in \{0, 1\}$ are defined by

$$f_o^j(i) = \begin{cases} f_g^j & \text{if } i \leq i_c, \\ 0 & \text{if } i > i_c, \end{cases}$$

where $i_c$ is the current tick index. This amounts to saying that all the fees were collected below the tick whose index is $i$, which clearly might not be true. However, since these values are only used for relative computations (as we shall see), this inaccuracy does not modify the results that we are interested in. Hence, the variables $f_o^0(i)$ and $f_o^1(i)$ do not have to be deemed to be real amounts of collected fees but rather auxiliary values that are needed for the computation of the fees that each position has earned. Observe also that an arbitrary definition as the previous one for the initial values of $f_o^0(i)$ and $f_o^1(i)$ is needed, since when the new tick index $i$ is initialized, there is no way to determine the amount of fees that correspond to each of the intervals $[0, t)$ and $[t, +\infty)$, where $t$ is the tick whose index is $i$.

## 5.5.2 Fees

As we mentioned at the beginning of this section, Uniswap v3 charges a trading fee on the incoming token for each trade that is made. The smart contract of Uniswap v3 also allows setting a *protocol fee*, which is a fraction of each trading fee that is kept aside for the protocol and hence is not given to the liquidity providers. Both the percentages of the trading fee and the protocol fee are set and fixed when a new liquidity pool is created. The percentage of the protocol fee is zero by default but can be defined to be any number among a permitted set of values. For simplicity, from now on, we will assume that the protocol fee is zero. This simplification will not affect the understanding of how Uniswap v3 works since the only consequence when the protocol fee is not zero is that the amount of trading fees changes, but the way they are collected and tracked in the different variables and paid to liquidity providers remains unmodified.

In order to compute the amount of fees that a certain position collects, we need to introduce the concepts of fees from above, fees from below, and fees within a range.

**Fees from above and below.** Given a tick $t$ with index $i$, and given $j \in \{0, 1\}$, we want to compute the fees in token $j$ that were collected (per unit of liquidity) from *above* and from *below* tick $t$, that is, in the intervals $[t, +\infty)$ and $[0, t)$, respectively. The fees from above the tick whose index is $i$ will be denoted by $f_a^j(i)$, and the fees from below the tick whose index is $i$ will be denoted by $f_b^j(i)$.

Let $j \in \{0, 1\}$. Recall that if $t$ is an initialized tick and $i$ is its index, then $f_o^j(i)$ represents the amount of fees in token $j$ (per unit of liquidity) that were collected in the interval $[0, t)$ if $i_c \geq i$, or in the interval $[t, +\infty)$ if $i_c < i$. Hence, we define

$$f_a^j(i) = \begin{cases} f_g^j - f_o^j(i) & \text{if } i_c \geq i \\ f_o^j(i) & \text{if } i_c < i \end{cases}$$

and

$$f_b^j(i) = \begin{cases} f_o^j(i) & \text{if } i_c \geq i, \\ f_g^j - f_o^j(i) & \text{if } i_c < i. \end{cases}$$

**Example 5.5.** Consider a Uniswap v3 pool with tokens $X$ and $Y$. Let $t_1$, $t_2$, and $t_3$ be ticks such that $t_1 < t_2 < t_3$. In Table 5-3, we show how the variables $f_g^j, f_o^j, f_a^j$, and $f_b^j$ $(j \in \{0,1\})$ are updated when some hypothetical trades are executed and the price moves. In order to focus on these variables, we take as inputs the price movements and the collected fees. The values of the variables that are updated in each step are highlighted in bold type.

***Table 5-3.*** *Example on how the variables $f_g, f_o^j, f_a^j$, and $f_b^j$ (for $j \in \{0, 1\}$) are updated*

| Input data | | Global fee | | Fee variables at tick $t_2$ | | | | | |
| --- | --- | --- | --- | --- | --- | --- | --- | --- | --- |
| Price movement | Collected fee | $f_g^0$ | $f_g^1$ | $f_o^0$ | $f_a^0$ | $f_b^0$ | $f_o^1$ | $f_a^1$ | $f_b^1$ |
| Initial price: $t_1$ | — | 0 | 0 | 0 | 0 | 0 | 0 | 0 | 0 |
| $t_1 \rightarrow t_2$ | 10 $Y$ | 0 | **10** | 0 | 0 | 0 | 0 | 0 | **10** |
| $t_2$ tick cross | — | 0 | 10 | 0 | 0 | 0 | **10** | 0 | 10 |
| $t_2 \rightarrow t_3$ | 15 $Y$ | 0 | **25** | 0 | 0 | 0 | 10 | **15** | 10 |
| $t_2 \leftarrow t_3$ | 4 $X$ | **4** | 25 | 0 | **4** | 0 | 10 | 15 | 10 |
| $t_2$ tick cross | — | 4 | 25 | **4** | 4 | 0 | **15** | 15 | 10 |
| $t_1 \leftarrow t_2$ | 3 $X$ | **7** | 25 | 4 | 4 | **3** | 15 | 15 | 10 |

It is important to observe that when a trade is performed and a fee is collected, the global fee variables $f_g^0$ and $f_g^1$ need to be updated–actually, just only one of them is updated because the fee is collected in only one of the two pool tokens. In consequence, the values of the variables $f_a^j$ and $f_b^j, j \in \{0,1\}$, might change as well. In addition, the values of the variables

$f_o^0$ and $f_o^1$ remain unmodified when a trade is performed (if no initialized ticks are crossed) since these variables represent the fees collected outside the price interval we are currently in, and clearly, the amounts of fees collected outside the current interval do not change when a trade is performed inside this interval.

On the other hand, when a tick is crossed, the variables $f_o^0$ and $f_o^1$ need to be updated, as we explained in the previous subsection. Observe that the values of the variables $f_a^j$ and $f_b^j, j \in \{0,1\}$, remain the same when a tick is crossed since the change in the values of the variables $f_o^0$ and $f_o^1$ is compensated by the modification in the formulae that define $f_a^j$ and $f_b^j, j \in \{0,1\}$.

In Table 5-3, we show how all this occurs. We have to mention, though, that actually, the variables $f_a^j$ and $f_b^j, j \in \{0,1\}$, are neither tracked nor updated in the Uniswap v3 smart contract since there is no need to do that. They are just auxiliary variables that are used to make easier the computation of the fees earned by a position. Nevertheless, we think that it is worth including in this example how these variables change in order to better understand how the Uniswap v3 smart contract works.

**Fees within a range.** Let $j \in \{0, 1\}$ and let $l$ and $u$ be ticks with $l < u$. Let $i_l$ be the tick index of $l$ and let $i_u$ be the tick index of $u$. We define $f_r^j(l,u)$ by

$$f_r^j(l,u) = f_g^j - f_b^j(i_l) - f_a^j(i_u).$$

Note that $f_r^j(l,u)$ represents the fees in token $j$ that were collected (per unit of liquidity) within the price range $[l, u)$. It is important to observe that since $f_r^j$ is defined using $f_a^j$ and $f_b^j$, which are in turn defined in terms of $f_o^j$, as it occurred with $f_g^j$, these variables cannot be used to obtain the real values of what they represent. However, they are useful to compute the amount of fees that a position has collected, as we will see next.

In Uniswap v3, a position consists of an address (from which the funds come) and two tick indexes that define the boundaries of the price interval in which liquidity is provided. When a liquidity provider creates a new position depositing a certain amount of liquidity in a price interval $[l, u]$, the smart contract of Uniswap v3 associates to that position–that is, the 3-tuple given by an address, the tick index of $l$, and the tick index of $u$–not only the liquidity parameter $L$ but also the values of $f_r^0(l,u)$ and $f_r^1(l,u)$ at the moment the deposit is made. In the code of the smart contract of Uniswap v3, these last two values are stored in two variables that are called feeGrowthInside0LastX128 and feeGrowthInside1LastX128, respectively, as we can see in the code given in Listing 5-1.[1]

***Listing 5-1.*** First part of the Position class of the smart contract of Uniswap v3

```
/// @title Position
/// @notice Positions represent an owner address'
  ↪liquidity between a lower and upper tick boundary
/// @dev Positions store additional state for tracking
  ↪fees owed to the position
library Position {
    // info stored for each user's position
    struct Info {
        // the amount of liquidity owned by this
  ↪position
        uint128 liquidity;
        // fee growth per unit of liquidity as of the
  ↪last update to liquidity or fees owed
        uint256 feeGrowthInside0LastX128;
        uint256 feeGrowthInside1LastX128;
```

---

[1] This code can be found in https://github.com/Uniswap/v3-core/blob/main/contracts/libraries/Position.sol

For simplicity, we will denote the initial values of $f_r^0(l,u)$ and $f_r^1(l,u)$ at the moment the deposit is made by $F_0$ and $F_1$. When the liquidity provider wants to redeem their fees, the smart contract takes the updated amounts $f_r^0(l,u)$ and $f_r^1(l,u)$ and gives to the liquidity provider an amount $L(f_r^0(l,u)-F_0)$ of token 0 and an amount $L(f_r^1(l,u)-F_1)$ of token 1. Then, the protocol updates the variables feeGrowthInside0LastX128 and feeGrowthInside1LastX128 as $f_r^0(l,u)$ and $f_r^1(l,u)$, respectively, so as to be able to compute the amounts of fees that the position collects from this time on.

We previously mentioned that the variables $f_o^j, f_a^j$, and $f_b^j$ cannot be used to determine the exact amount of collected fees in the intervals they track, since when a new tick index $i$ is initialized, the variable $f_o^j(i)$ is arbitrarily defined in a certain way that does not measure the amount of fees that were collected before. However, after that initial state, and since the variable $f_g^j$ is continuously incremented with the new fees that are charged, the variables $f_o^j(i), f_a^j(i)$, and $f_b^j(i)$ are also updated with those increments on fees (when the fees are collected in the corresponding intervals, of course). Therefore, the differences $f_r^0(l,u)-F_0$ and $f_r^1(l,u)-F_1$ do indeed track the increments in the amounts of fees (per unit of liquidity) since the last time the liquidity provider redeemed their share of fees.

## 5.5.3 Trades

In Uniswap v3, trades are performed in a similar way as in Uniswap v2, following a constant product formula. However, in order to perform trades in Uniswap v3, we need to take into account the list of initialized ticks, since the liquidity parameter might change when the price crosses an initialized tick.

We will see now how this works. Let $\phi$ be the trading fee and let $p$ be the current price, which is represented by the tick index $i_c$. Let $\mathcal{I}$ be the set of initialized tick indexes and let

$$i_u = \min\{i \in \mathcal{I} \mid i > i_c\}$$

and

$$i_l = \max\{i \in \mathcal{I} \mid i \leq i_c\}.$$

Informally, $i_u$ and $i_l$ are the initialized tick indexes that are the nearest to $i_c$ that are located to the right and to the left of $i_c$, respectively. Observe that $i_c \in [i_l, i_u)$ and that the interval $(i_l, i_u)$ does not contain any initialized tick indexes. Thus, the liquidity parameter does not change within the interval $[i_l, i_u]$.

Let $t_l$ and $t_u$ be the ticks whose indexes are $i_l$ and $i_u$, respectively. Note that $p \in [t_l, t_u)$ since $i_c \in [i_l, i_u)$. When a trade is going to be made, the protocol checks if the trade can be made within the current tick interval $[t_l, t_u]$; that is, the protocol checks if the liquidity of the interval $[t_l, t_u]$ is enough to perform the trade. If this is the case, the trade is executed, and the corresponding fees are collected, as we explained in previous sections. On the other hand, if the liquidity on that interval is not enough to perform the whole trade, then a portion of the trade is performed until the price reaches the boundary of the interval $[t_l, t_u]$, and then the trade continues on the next interval (to the right or the left depending on which boundary was reached) with the remaining part of the trade. We mention that the intervals considered are formed by consecutive initialized ticks, since ticks that are not initialized might be ignored.

We will now explain the process of the previous paragraph in greater detail. To this end, let $L$ be the liquidity parameter of the interval $[t_l, t_u]$. Recall that the virtual balances are given by

$$x_v = \frac{L}{\sqrt{p}} \quad \text{and} \quad y_v = L\sqrt{p}.$$

We will analyze four cases.

<u>Case 1</u>: The trader wants to obtain an amount $a$ of token $X$.

Let $x_r$ be the real balance of token $X$ (within the interval $[t_l, t_u]$). From Table 5-1, we obtain that

$$x_r = L\left(\frac{1}{\sqrt{p}} - \frac{1}{\sqrt{t_u}}\right).$$

Thus, if $a \le x_r$, the trade can be performed within the interval $[t_l, t_u]$. In this case, the trade is executed, and since the virtual balance of token $X$ after the trade is $x_v - a$, the price is updated as

$$p' = \left(\frac{L}{x_v - a}\right)^2.$$

On the other hand, if $a > x_r$, then the trade cannot be carried out completely within the interval $[t_l, t_u]$. In this case, a first step of the trade is computed, where an (output) amount $x_r$ of token $X$ is traded, the necessary (input) amount of token $Y$ is computed and deposited, and a certain fee on the token $Y$ is collected. Note that the virtual balance of token $X$ after this first step is

$$x_v - x_r = \frac{L}{\sqrt{p}} - L\left(\frac{1}{\sqrt{p}} - \frac{1}{\sqrt{t_u}}\right) = \frac{L}{\sqrt{t_u}}$$

and thus, the updated price would be equal to $t_u$, as expected. Now, since $a > x_r$, there is a remaining amount of token $X$ to be included in the trade (the remaining amount is $a - x_r$). Hence, tick $t_u$ needs to be crossed, and the trade continues on the next interval, which is $[t_u, t']$, where $t'$ is the smallest initialized tick that is greater than $t_u$. When tick $t_u$ is crossed, the variables $i_c, L_{tot}, f_o^0(i_u)$, and $f_o^1(i_u)$ are updated. After that, the remaining amount of token $X$ is traded considering the interval $[t_u, t']$.

Clearly, the same considerations explained previously apply to the remaining part of the trade. If the real balance of token $X$ within the interval $[t_u, t']$ is enough to perform the rest of the trade, then the trade is completed, the necessary (input) amount of token $Y$ is computed and deposited, and the corresponding fee on token $Y$ is collected. On the other hand, if the trade cannot be completed within the interval $[t_u, t']$, another fraction of the trade is computed, and the trade continues on the next interval.

Case 2: The trader wants to deposit an amount $a$ of token $X$ in order to obtain an amount of token $Y$.

Let $a' = (1 - \phi)a$. As in the previous case, let $x_r$ be the real balance of token $X$ (within the interval $[t_l, t_u]$) and let $x_{max}$ be the maximum possible value of the real balance of token $X$ in the interval $[t_l, t_u]$. From Table 5-1, we obtain that

$$x_r = L\left( \frac{1}{\sqrt{p}} - \frac{1}{\sqrt{t_u}} \right) \quad \text{and} \quad x_{max} = L\left( \frac{1}{\sqrt{t_l}} - \frac{1}{\sqrt{t_u}} \right).$$

Thus, if $x_r + a' \le x_{max}$, the trade can be completed within the interval $[t_l, t_u]$. In this case, a fee $\phi a$ is charged and the remaining amount $a'$ of token $X$ is traded, and since the virtual balance of token $X$ after the trade is $x_v + a'$, the price is updated as

$$p' = \left( \frac{L}{x_v + a'} \right)^2.$$

On the other hand, if $x_r + a' > x_{max}$, then the trade cannot be completed within the interval $[t_l, t_u]$. As in the previous case, in this situation, a first step of the trade is computed, where the amount of token $X$ that is traded is

$$x_{max} - x_r = L\left( \frac{1}{\sqrt{t_l}} - \frac{1}{\sqrt{t_u}} \right) - L\left( \frac{1}{\sqrt{p}} - \frac{1}{\sqrt{t_u}} \right) = L\left( \frac{1}{\sqrt{t_l}} - \frac{1}{\sqrt{p}} \right).$$

Note that $a' > x_{\max} - x_r$ and that the virtual balance of token $X$ after this first step is

$$x_v + x_{\max} - x_r \quad = \frac{L}{\sqrt{p}} + L\left(\frac{1}{\sqrt{t_l}} - \frac{1}{\sqrt{t_u}}\right) - L\left(\frac{1}{\sqrt{p}} - \frac{1}{\sqrt{t_u}}\right) =$$

$$= \frac{L}{\sqrt{t_l}}$$

and thus, the updated price would be equal to $t_l$, as expected. And since $a' > x_{\max} - x_r$, there is a remaining amount of token $X$ to be included in the trade. Observe that since the amount of token $X$ that was traded in this first step is $x_{\max} - x_r$, the amount of token $X$ that the trader had to use for this step is $\frac{1}{1-\phi}(x_{\max} - x_r)$. Therefore, the remaining amount of token $X$ that is available for the next step of the trade (without fees being charged yet) is

$$a - \frac{x_{\max} - x_r}{1-\phi}$$

or equivalently,

$$\frac{a'}{1-\phi} - \frac{x_{\max} - x_r}{1-\phi} = \frac{1}{1-\phi}\left(a' - (x_{\max} - x_r)\right).$$

Note also that the trading fee that was charged in the first step is $\frac{\phi}{1-\phi}(x_{\max} - x_r)$.

After the first step of the trade is completed, tick $t_l$ is crossed, and the trade continues in the interval $[t', t_l]$, where $t'$ is the greatest initialized tick that is smaller than $t_l$. As in Case 1, when tick $t_l$ is crossed, the variables $i_c, L_{tot}, f_o^0(i_l)$, and $f_o^1(i_l)$ are updated.

It is important to mention that the rest of the trade is executed taking into account the previous discussion regarding this case and considering the interval $[t', t_l]$. Specifically, if the remaining part of the trade can be completed within the interval $[t', t_l]$, then this remaining part is performed, and the whole trade is completed. On the other hand, if the remaining part of the trade cannot be completed within the interval $[t', t_l]$, a suitable portion of the trade is executed in that interval, and the trade continues in the next one (to the left).

Case 3: The trader wants to obtain an amount $b$ of token $Y$.

As in Case 1, let $y_r$ be the real balance of token $Y$ (within the interval $[t_l, t_u]$). From Table 5-1, we know that

$$y_r = L\left(\sqrt{p} - \sqrt{t_l}\right).$$

Thus, if $b \leq y_r$, the trade can be completed within the interval $[t_l, t_u]$. In this case, the trade is executed, and the price is updated as

$$p' = \left(\frac{y_v - b}{L}\right)^2.$$

On the other hand, if $b > y_r$, then the trade cannot be completed within the interval $[t_l, t_u]$. Hence, the first step of the trade is computed, with an (output) amount $y_r$ of token $Y$ and the corresponding (input) amount of token $X$. Note that the fee is charged on token $X$. Clearly, the virtual balance of token $Y$ after this first step is

$$y_v - y_r = L\sqrt{p} - L\left(\sqrt{p} - \sqrt{t_l}\right) = L\sqrt{t_l}$$

and thus, the price is $t_l$, as expected. As there is a remaining amount of token $Y$ to be included in the trade (since $b > y_r$), tick $t_l$ is crossed, and the trade continues in the interval $[t', t_l]$, where $t'$ is the greatest initialized tick that is smaller than $t_l$. Note that the remaining amount of token $Y$ to be traded is $b - y_r$. As in Case 2, the variables $i_c, L_{tot}, f_o^0(i_l)$, and $f_o^1(i_l)$ are updated when tick $t_l$ is crossed.

Again, the same considerations discussed previously apply for the remaining part of the trade with respect to the interval $[t', t_l]$, in a similar way as in the previous cases.

Case 4: The trader wants to deposit an amount $b$ of token $Y$ in order to obtain an amount of token $X$.

We proceed in a similar way as in Case 2. Let $b' = (1 - \phi)b$, let $y_r$ be the real balance of token $Y$ (within the interval $[t_l, t_u]$), and let $y_{max}$ be the maximum possible value of the real balance of token $Y$ in the interval $[t_l, t_u]$. From Table 5-1, we know that

$$y_r = L\left(\sqrt{p} - \sqrt{t_l}\right) \quad \text{and} \quad y_{max} = L\left(\sqrt{t_u} - \sqrt{t_l}\right).$$

Thus, if $y_r + b' \leq y_{max}$, the trade can be completed within the interval $[t_l, t_u]$. In this case, a fee $\phi b$ is charged, the remaining amount $b'$ of token $Y$ is traded, and the price is updated as

$$p' = \left(\frac{y_v + b'}{L}\right)^2.$$

On the other hand, if $y_r + b' > y_{max}$, then the trade cannot be completed within the interval $[t_l, t_u]$. Thus, a first step of the trade is computed, where the amount of token $Y$ that is traded is

$$y_{max} - y_r = L\left(\sqrt{t_u} - \sqrt{t_l}\right) - L\left(\sqrt{p} - \sqrt{t_l}\right) = L\left(\sqrt{t_u} - \sqrt{p}\right)$$

Note that $b' > y_{max} - y_r$ and that the virtual balance of token $Y$ after this first step is

$$y_v + y_{max} - y_r = L\sqrt{p} + L\left(\sqrt{t_u} - \sqrt{t_l}\right) - L\left(\sqrt{p} - \sqrt{t_l}\right) = L\sqrt{t_u}$$

and thus, the updated price is $t_u$, as expected. Since $b' > y_{max} - y_r$, there is a remaining amount of token $Y$ to be traded. As in Case 2, since the amount of token $Y$ that was traded in the first step is $y_{max} - y_r$, the amount of token $Y$ that the trader had to use for this step is $\dfrac{1}{1-\phi}(y_{max} - y_r)$. Hence, the remaining amount of token $Y$ that is available for the next step of the trade (without fees being charged yet) is

$$b - \frac{y_{max} - y_r}{1-\phi}$$

or equivalently,

$$\frac{b'}{1-\phi} - \frac{y_{max} - y_r}{1-\phi} = \frac{1}{1-\phi}\left(b' - \left(y_{max} - y_r\right)\right).$$

Note also that the trading fee that was charged in the first step is $\dfrac{\phi}{1-\phi}(y_{max} - y_r)$.

After the first step of the trade is completed, tick $t_u$ is crossed, and the trade continues in the interval $[t_u, t']$, where $t'$ is the smallest initialized tick that is greater than $t_u$. Observe that, as in the previous cases, when tick $t_u$ is crossed, the variables $i_c, L_{tot}, f_o^0(i_u)$, and $f_o^1(i_u)$ are updated.

It is worth mentioning that the same considerations discussed previously apply to the remaining part of the trade with respect to the interval $[t_u, t']$. Explicitly, if the remaining part of the trade can be completed within the interval $[t_u, t']$, then this remaining part is executed, and the whole trade is completed. On the other hand, if the remaining part of the trade cannot be completed within the interval $[t_u, t']$, a suitable portion of the trade is performed in that interval, and the trade continues in the next one.

We will make here an additional remark. In all four cases, when the trade cannot be completed within the current interval, we have implicitly assumed that a next (or previous) initialized tick $t'$ exists, which defines the next interval $[t_u, t']$ or the previous interval $[t', t_l]$ according to the situation. Although this is usually the case, it might happen that such tick $t'$ does not exist. This implies that there is no liquidity above $t_u$ (or below $t_l$), and hence, the trade cannot continue on the next (or previous) interval. Thus, the algorithm will stop, and the trade will have been only partially fulfilled.

## 5.5.4 The swap Function

The trading process that we explained before is carried out by the swap function.[2] This function makes use of the computeSwapStep function of the SwapMath library,[3] which is included in Listing 5-2. The computeSwapStep function computes and returns the parameters of a trade up to the next initialized tick, as we can see from the following code. In order to understand the code, it is important to know the roles of the boolean variables zeroForOne and exactIn. The boolean variable zeroForOne is 1 if the price is moving downward (token 0 is going into the pool, and token 1 is going out of the pool) and 0 if the price is moving upward (token 1 is going into the pool, and token 0 is going out of the pool). On the other hand, the boolean variable exactIn is 1 if we are given the amount of either token 0 or token 1 that goes into the pool as a parameter of the trade, that is, if the trader wants to deposit a certain amount of either token and we need to compute the amount of the other token that they will obtain. And clearly, the boolean variable exactIn is 0 if we are given the amount (of either token) that goes out of the pool (and thus, we will need to compute the amount that needs to be deposited).

---

[2] https://github.com/Uniswap/v3-core/blob/main/contracts/
UniswapV3Pool.sol
[3] https://github.com/Uniswap/v3-core/blob/main/contracts/libraries/
SwapMath.sol

***Listing 5-2.*** Function `computeSwapStep` of the smart contract of Uniswap v3

```
/// @title Computes the result of a swap within ticks
/// @notice Contains methods for computing the result
  ↪of a swap within a single tick price range, i.e., a
  ↪single tick.
library SwapMath {
    /// @notice Computes the result of swapping some
  ↪amount in, or amount out, given the parameters of
  ↪the swap
    /// @dev The fee, plus the amount in, will never
  ↪exceed the amount remaining if the swap's
  ↪ `amountSpecified` is positive
    /// @param sqrtRatioCurrentX96 The current sqrt
  ↪price of the pool
    /// @param sqrtRatioTargetX96 The price that cannot
  ↪be exceeded, from which the direction of the swap
  ↪is inferred
    /// @param liquidity The usable liquidity
    /// @param amountRemaining How much input or output
  ↪amount is remaining to be swapped in/out
    /// @param feePips The fee taken from the input
  ↪amount, expressed in hundredths of a bip
    /// @return sqrtRatioNextX96 The price after
  ↪swapping the amount in/out, not to exceed the price
  ↪target
    /// @return amountIn The amount to be swapped in,
  ↪of either token0 or token1, based on the direction
  ↪of the swap
    /// @return amountOut The amount to be received, of
  ↪either token0 or token1, based on the direction of
```

```
↪the swap
  /// @return feeAmount The amount of input that will
↪be taken as a fee
  function computeSwapStep(
      uint160 sqrtRatioCurrentX96,
      uint160 sqrtRatioTargetX96,
      uint128 liquidity,
      int256 amountRemaining,
      uint24 feePips
  )
      internal
      pure
      returns (
          uint160 sqrtRatioNextX96,
          uint256 amountIn,
          uint256 amountOut,
          uint256 feeAmount
      )
  {
      bool zeroForOne = sqrtRatioCurrentX96 >=
↪sqrtRatioTargetX96;
      bool exactIn = amountRemaining >= 0;

      if (exactIn) {
          uint256 amountRemainingLessFee =
↪FullMath.mulDiv(uint256(amountRemaining), 1e6 -
↪feePips, 1e6);
          amountIn = zeroForOne
              ?
↪SqrtPriceMath.getAmount0Delta(sqrtRatioTargetX96,
↪sqrtRatioCurrentX96, liquidity, true)
              :
```

```
↪SqrtPriceMath.getAmount1Delta(sqrtRatioCurrentX96,
↪sqrtRatioTargetX96, liquidity, true);
        if (amountRemainingLessFee >= amountIn)
↪sqrtRatioNextX96 = sqrtRatioTargetX96;
        else
            sqrtRatioNextX96 =
↪SqrtPriceMath.getNextSqrtPriceFromInput(
                sqrtRatioCurrentX96,
                liquidity,
                amountRemainingLessFee,
                zeroForOne
            );
    } else {
        amountOut = zeroForOne
            ?
↪SqrtPriceMath.getAmount1Delta(sqrtRatioTargetX96,
↪sqrtRatioCurrentX96, liquidity, false)
            :
↪SqrtPriceMath.getAmount0Delta(sqrtRatioCurrentX96,
↪sqrtRatioTargetX96, liquidity, false);
        if (uint256(-amountRemaining) >= amountOut)
↪sqrtRatioNextX96 = sqrtRatioTargetX96;
        else
            sqrtRatioNextX96 =
↪SqrtPriceMath.getNextSqrtPriceFromOutput(
                sqrtRatioCurrentX96,
                liquidity,
                uint256(-amountRemaining),
                zeroForOne
            );
    }
```

```
    bool max = sqrtRatioTargetX96 ==
↪sqrtRatioNextX96;

        // get the input/output amounts
        if (zeroForOne) {
            amountIn = max && exactIn
                ? amountIn
                :
↪SqrtPriceMath.getAmount0Delta(sqrtRatioNextX96,
↪sqrtRatioCurrentX96, liquidity, true);
            amountOut = max && !exactIn
                ? amountOut
                :
↪SqrtPriceMath.getAmount1Delta(sqrtRatioNextX96,
↪sqrtRatioCurrentX96, liquidity, false);
        } else {
            amountIn = max && exactIn
                ? amountIn
                :
↪SqrtPriceMath.getAmount1Delta(sqrtRatioCurrentX96,
↪sqrtRatioNextX96, liquidity, true);
            amountOut = max && !exactIn
                ? amountOut
                :
↪SqrtPriceMath.getAmount0Delta(sqrtRatioCurrentX96,
↪sqrtRatioNextX96, liquidity, false);
        }

        // cap the output amount to not exceed the
↪remaining output amount
        if (!exactIn && amountOut >
```

```
↪uint256(-amountRemaining)) {
        amountOut = uint256(-amountRemaining);
    }

    if (exactIn && sqrtRatioNextX96 !=
↪sqrtRatioTargetX96) {
        // we didn't reach the target, so take the
↪remainder of the maximum input as fee
        feeAmount = uint256(amountRemaining) -
↪amountIn;
    } else {
        feeAmount =
↪FullMath.mulDivRoundingUp(amountIn, feePips, 1e6 -
↪feePips);
    }
  }
}
```

The swap function also uses three structs. Two of them are called SwapState and StepComputations and are given in Listing 5-3. The SwapState object tracks the cumulative state of the trade, that is, the parameters of the trade after each step of it, together with the total amounts of each token that have been computed after each step and the remaining amount to be traded. On the other hand, the StepComputations object collects the parameters of the current step of the trade.

*Listing 5-3.* SwapState and StepComputations structs of the smart contract of Uniswap v3

```
// the top level state of the swap, the results of
↪which are recorded in storage at the end
  struct SwapState {
      // the amount remaining to be swapped in/out of
```

```
↪the input/output asset
      int256 amountSpecifiedRemaining;
      // the amount already swapped out/in of the
↪output/input asset
      int256 amountCalculated;
      // current sqrt(price)
      uint160 sqrtPriceX96;
      // the tick associated with the current price
      int24 tick;
      // the global fee growth of the input token
      uint256 feeGrowthGlobalX128;
      // amount of input token paid as protocol fee
      uint128 protocolFee;
      // the current liquidity in range
      uint128 liquidity;
  }

  struct StepComputations {
      // the price at the beginning of the step
      uint160 sqrtPriceStartX96;
      // the next tick to swap to from the current
↪tick in the swap direction
      int24 tickNext;
      // whether tickNext is initialized or not
      bool initialized;
      // sqrt(price) for the next tick (1/0)
      uint160 sqrtPriceNextX96;
      // how much is being swapped in in this step
      uint256 amountIn;
      // how much is being swapped out
      uint256 amountOut;
```

```
    // how much fee is being paid in
    uint256 feeAmount;
}
```

Now we are ready to analyze the swap function, which is given in Listing 5-4.

***Listing 5-4.*** swap function of the smart contract of Uniswap v3

```
/// @inheritdoc IUniswapV3PoolActions
/// @notice Swap token0 for token1, or token1 for
↪token0
/// @dev The caller of this method receives a
↪callback in the form of
↪IUniswapV3SwapCallback#uniswapV3SwapCallback
/// @param recipient The address to receive the
↪output of the swap
/// @param zeroForOne The direction of the swap,
↪true for token0 to token1, false for token1 to
↪token0
/// @param amountSpecified The amount of the swap,
↪which implicitly configures the swap as exact input
↪(positive), or exact output (negative)
/// @param sqrtPriceLimitX96 The Q64.96 sqrt price
↪limit. If zero for one, the price cannot be less
↪than this
/// value after the swap. If one for zero, the
↪price cannot be greater than this value after the
↪swap
/// @param data Any data to be passed through to
↪the callback
/// @return amount0 The delta of the balance of
```

*↪token0 of the pool, exact when negative, minimum*
*↪when positive*
*/// @return amount1 The delta of the balance of*
*↪token1 of the pool, exact when negative, minimum*
*↪when positive*

```
  function swap(
      address recipient,
      bool zeroForOne,
      int256 amountSpecified,
      uint160 sqrtPriceLimitX96,
      bytes calldata data
  ) external override noDelegateCall returns (int256
↪amount0, int256 amount1) {
      require(amountSpecified != 0, 'AS');

      Slot0 memory slot0Start = slot0;

      require(slot0Start.unlocked, 'LOK');
      require(
          zeroForOne
              ? sqrtPriceLimitX96 <
↪slot0Start.sqrtPriceX96 && sqrtPriceLimitX96 >
↪TickMath.MIN_SQRT_RATIO
              : sqrtPriceLimitX96 >
↪slot0Start.sqrtPriceX96 && sqrtPriceLimitX96 <
↪TickMath.MAX_SQRT_RATIO,
          'SPL'
      );

      slot0.unlocked = false;
```

```
    SwapCache memory cache =
        SwapCache({
            liquidityStart: liquidity,
            blockTimestamp: _blockTimestamp(),
            feeProtocol: zeroForOne ?
↪(slot0Start.feeProtocol % 16) :
↪(slot0Start.feeProtocol >> 4),
            secondsPerLiquidityCumulativeX128: 0,
            tickCumulative: 0,
            computedLatestObservation: false
        });

    bool exactInput = amountSpecified > 0;

    SwapState memory state =
        SwapState({
            amountSpecifiedRemaining:
↪amountSpecified,
            amountCalculated: 0,
            sqrtPriceX96: slot0Start.sqrtPriceX96,
            tick: slot0Start.tick,
            feeGrowthGlobalX128: zeroForOne ?
↪feeGrowthGlobal0X128 : feeGrowthGlobal1X128,
            protocolFee: 0,
            liquidity: cache.liquidityStart
        });

    // continue swapping as long as we haven't used
↪the entire input/output and haven't reached the
↪price limit
    while (state.amountSpecifiedRemaining != 0 &&
```

```
↪state.sqrtPriceX96 != sqrtPriceLimitX96) {
        StepComputations memory step;

        step.sqrtPriceStartX96 =
↪state.sqrtPriceX96;

        (step.tickNext, step.initialized) =
↪tickBitmap.nextInitializedTickWithinOneWord(
            state.tick,
            tickSpacing,
            zeroForOne
        );

        // ensure that we do not overshoot the
↪min/max tick, as the tick bitmap is not aware of
↪these bounds
        if (step.tickNext < TickMath.MIN_TICK) {
            step.tickNext = TickMath.MIN_TICK;
        } else if (step.tickNext >
↪TickMath.MAX_TICK) {
            step.tickNext = TickMath.MAX_TICK;
        }

        // get the price for the next tick
        step.sqrtPriceNextX96 =
↪TickMath.getSqrtRatioAtTick(step.tickNext);

        // compute values to swap to the target
↪tick, price limit, or point where input/output
↪amount is exhausted
        (state.sqrtPriceX96, step.amountIn,
↪step.amountOut, step.feeAmount) =
↪SwapMath.computeSwapStep(
            state.sqrtPriceX96,
```

```
            (zeroForOne ? step.sqrtPriceNextX96 <
↪sqrtPriceLimitX96 : step.sqrtPriceNextX96 >
↪sqrtPriceLimitX96)
                    ? sqrtPriceLimitX96
                    : step.sqrtPriceNextX96,
            state.liquidity,
            state.amountSpecifiedRemaining,
            fee
        );

        if (exactInput) {
            state.amountSpecifiedRemaining -=
↪(step.amountIn + step.feeAmount).toInt256();
            state.amountCalculated =
↪state.amountCalculated.sub(step.amountOut.toInt256());
        } else {
            state.amountSpecifiedRemaining +=
↪step.amountOut.toInt256();
            state.amountCalculated =
↪state.amountCalculated.add((step.amountIn +
↪step.feeAmount).toInt256());
        }

        // if the protocol fee is on, calculate how
↪much is owed, decrement feeAmount, and increment
↪protocolFee
        if (cache.feeProtocol > 0) {
            uint256 delta = step.feeAmount /
↪cache.feeProtocol;
            step.feeAmount -= delta;
            state.protocolFee += uint128(delta);
        }
```

```
            // update global fee tracker
            if (state.liquidity > 0)
                state.feeGrowthGlobalX128 +=
↪FullMath.mulDiv(step.feeAmount, FixedPoint128.Q128,
↪state.liquidity);

            // shift tick if we reached the next price
            if (state.sqrtPriceX96 ==
↪step.sqrtPriceNextX96) {
                // if the tick is initialized, run the
↪tick transition
                if (step.initialized) {
                // check for the placeholder value,
↪which we replace with the actual value the first
↪time the swap
                // crosses an initialized tick
                if
↪(!cache.computedLatestObservation) {
                        (cache.tickCumulative,
↪cache.secondsPerLiquidityCumulativeX128) =
↪observations.observeSingle(
                        cache.blockTimestamp,
                        0,
                        slot0Start.tick,

↪slot0Start.observationIndex,
                        cache.liquidityStart,

↪slot0Start.observationCardinality
                    );
                    cache.computedLatestObservation
↪= true;
                }
```

```
                int128 liquidityNet =
                    ticks.cross(
                        step.tickNext,
                        (zeroForOne ?
↪state.feeGrowthGlobalX128 : feeGrowthGlobal0X128),
                        (zeroForOne ?
↪feeGrowthGlobal1X128 : state.feeGrowthGlobalX128),
↪cache.secondsPerLiquidityCumulativeX128,
                        cache.tickCumulative,
                        cache.blockTimestamp
                    );
                // if we're moving leftward, we
↪interpret liquidityNet as the opposite sign
                // safe because liquidityNet cannot
↪be type(int128).min
                if (zeroForOne) liquidityNet =
↪-liquidityNet;
                state.liquidity =
↪LiquidityMath.addDelta(state.liquidity, liquidityNet);
            }

            state.tick = zeroForOne ? step.tickNext
↪- 1 : step.tickNext;
        } else if (state.sqrtPriceX96 !=
↪step.sqrtPriceStartX96) {
            // recompute unless we're on a lower
↪tick boundary (i.e. already transitioned ticks),
↪and haven't moved
            state.tick =
↪TickMath.getTickAtSqrtRatio(state.sqrtPriceX96);
        }
    }
```

```
    // update tick and write an oracle entry if the
↪tick change
    if (state.tick != slot0Start.tick) {
        (uint16 observationIndex, uint16
↪observationCardinality) =
            observations.write(
                slot0Start.observationIndex,
                cache.blockTimestamp,
                slot0Start.tick,
                cache.liquidityStart,
                slot0Start.observationCardinality,
↪slot0Start.observationCardinalityNext
            );
        (slot0.sqrtPriceX96, slot0.tick,
↪slot0.observationIndex,
↪slot0.observationCardinality) = (
            state.sqrtPriceX96,
            state.tick,
            observationIndex,
            observationCardinality
        );
    } else {
        // otherwise just update the price
        slot0.sqrtPriceX96 = state.sqrtPriceX96;
    }

    // update liquidity if it changed
    if (cache.liquidityStart != state.liquidity)
↪liquidity = state.liquidity;

    // update fee growth global and, if necessary,
↪protocol fees
```

```
    // overflow is acceptable, protocol has to
↪withdraw before it hits type(uint128).max fees
    if (zeroForOne) {
        feeGrowthGlobal0X128 =
↪state.feeGrowthGlobalX128;
    if (state.protocolFee > 0)
↪protocolFees.token0 += state.protocolFee;
    } else {
        feeGrowthGlobal1X128 =
↪state.feeGrowthGlobalX128;
    if (state.protocolFee > 0)
↪protocolFees.token1 += state.protocolFee;
    }

    (amount0, amount1) = zeroForOne == exactInput
        ? (amountSpecified -
↪state.amountSpecifiedRemaining,
↪state.amountCalculated)
        : (state.amountCalculated, amountSpecified
↪- state.amountSpecifiedRemaining);

    // do the transfers and collect payment
    if (zeroForOne) {
        if (amount1 < 0)
↪TransferHelper.safeTransfer(token1, recipient,
↪uint256(-amount1));

        uint256 balance0Before = balance0();
↪IUniswapV3SwapCallback(msg.sender).
 uniswapV3SwapCallback(amount0
↪amount1, data);
```

```
↪require(balance0Before.add(uint256(amount0)) <=
↪balance0(), 'IIA');
        } else {
            if (amount0 < 0)
↪TransferHelper.safeTransfer(token0, recipient,
↪uint256(-amount0));

            uint256 balance1Before = balance1();

↪IUniswapV3SwapCallback(msg.sender).
uniswapV3SwapCallback(amount0
    ↪amount1, data);

↪require(balance1Before.add(uint256(amount1)) <=
↪balance1(), 'IIA');
        }

        emit Swap(msg.sender, recipient, amount0,
↪amount1, state.sqrtPriceX96, state.liquidity,
↪state.tick);
        slot0.unlocked = true;
    }
```

The main part of the previous code is the while loop shown in Listing 5-5 that makes steps of the trade one after another until there is no amount left to trade.

***Listing 5-5.*** while loop of the swap function of the smart contract of Uniswap v3

```
// continue swapping as long as we haven't used the
↪entire input/output and haven't reached the price
↪limit
        while (state.amountSpecifiedRemaining != 0 &&
↪state.sqrtPriceX96 != sqrtPriceLimitX96) {
```

In Listing 5-6, we can see how the computeSwapStep function is used within the while loop.

**Listing 5-6.**  Usage of the computeSwapStep function in the smart contract of Uniswap v3

```
// get the price for the next tick
        step.sqrtPriceNextX96 =
↳TickMath.getSqrtRatioAtTick(step.tickNext);

        // compute values to swap to the target
↳tick, price limit, or point where input/output
↳amount is exhausted
        (state.sqrtPriceX96, step.amountIn,
↳step.amountOut, step.feeAmount) =
↳SwapMath.computeSwapStep(
            state.sqrtPriceX96,
            (zeroForOne ? step.sqrtPriceNextX96 <
↳sqrtPriceLimitX96 : step.sqrtPriceNextX96 >
↳sqrtPriceLimitX96)
            ? sqrtPriceLimitX96
            : step.sqrtPriceNextX96,
            state.liquidity,
            state.amountSpecifiedRemaining,
            fee
        );
```

In Listing 5-7, we can also see how the fee amount that was computed using the computeSwapStep function is added to the global fee variable of the state object after dividing it by the liquidity parameter (recall that the global fee variable and the outside fee variable are computed per unit of liquidity).

***Listing 5-7.*** The computed fee is added to the global fee variable

```
// update global fee tracker
if (state.liquidity > 0)
        state.feeGrowthGlobalX128 +=
↪FullMath.mulDiv(step.feeAmount, FixedPoint128.Q128,
↪state.liquidity);
```

Then, before the end of the while loop, the liquidity parameter is updated in order to set it for the next step, as we can see in Listing 5-8.

***Listing 5-8.*** The liquidity parameter is updated when an initialized tick is crossed

```
// if we're moving leftward, we
↪interpret liquidityNet as the opposite sign
                // safe because liquidityNet cannot
↪be type(int128).min
                if (zeroForOne) liquidityNet =
↪-liquidityNet;

                state.liquidity =
↪LiquidityMath.addDelta(state.liquidity,
↪liquidityNet);
```

After all the steps are completed–and hence outside the while loop–the global fee variable is updated, as is shown in Listing 5-9.

***Listing 5-9.*** The global fee variable is updated

```
// update fee growth global and, if necessary,
↪protocol fees
        // overflow is acceptable, protocol has to
↪withdraw before it hits type(uint128).max fees
```

```
    if (zeroForOne) {
        feeGrowthGlobal0X128 =
↪state.feeGrowthGlobalX128;
        if (state.protocolFee > 0)
↪protocolFees.token0 += state.protocolFee;
    } else {
        feeGrowthGlobal1X128 =
↪state.feeGrowthGlobalX128;
        if (state.protocolFee > 0)
↪protocolFees.token1 += state.protocolFee;
    }
```

And, of course, the corresponding transfers of tokens are performed and logged, as we can see in Listing 5-10.

***Listing 5-10.*** The token tranfers are performed

```
    // do the transfers and collect payment
    if (zeroForOne) {
    if (amount1 < 0)
↪TransferHelper.safeTransfer(token1, recipient,
↪uint256(-amount1));

    uint256 balance0Before = balance0();

↪IUniswapV3SwapCallback(msg.sender).uniswapV3Swap
Callback(amount0
↪amount1, data);

↪require(balance0Before.add(uint256(amount0)) <=
↪balance0(), 'IIA');
    } else {
        if (amount0 < 0)
```

```
↪TransferHelper.safeTransfer(token0, recipient,
↪uint256(-amount0));

            uint256 balance1Before = balance1();

↪IUniswapV3SwapCallback(msg.sender).uniswapV3Swap
Callback(amount0
↪amount1, data);

↪require(balance1Before.add(uint256(amount1)) <=
↪balance1(), 'IIA');
        }

        emit Swap(msg.sender, recipient, amount0,
↪amount1, state.sqrtPriceX96, state.liquidity,
↪state.tick);
        slot0.unlocked = true;
```

## 5.5.5 Example

We will now give a detailed example to illustrate how the Uniswap v3 protocol works.

Consider a Uniswap v3 liquidity pool with tokens ETH and USDC, a fee of 0.3%, and a tick spacing equal to 1. Let $\phi = 0.003$. Suppose that the current price of ETH in terms of USDC is 3,800 and that exactly two liquidity providers, A and B, provided liquidity to the pool. Liquidity provider A chose to provide liquidity in the interval [3,600.005212, 3,999.742678] depositing 3 ETH and a suitable amount of USDC. Liquidity provider B decided to provide liquidity in the interval [3,899.823492, 4,099.761469] depositing 10 ETH. Observe that liquidity provider B only needs to deposit ETH since the current price of ETH is below the interval they chose.

From Equation 5.1, we obtain that the index of tick 3,600.005212 is 81,891. Let $i_1 = 81,891$ and let $t_1 = (1.0001)^{81891} \approx 3,600.005212$. In a similar way, let

$$i_2 = 82,691, \qquad t_2 = (1.0001)^{82691} \approx 3,899.823492,$$
$$i_3 = 82,944, \qquad t_3 = (1.0001)^{82944} \approx 3,999.742678,$$
$$i_4 = 83,191, \quad \text{and} \quad t_4 = (1.0001)^{83191} \approx 4,099.761469.$$

Thus, liquidity provider $A$ deposited liquidity in the range $[t_1, t_3]$, and liquidity provider $B$ deposited liquidity in the interval $[t_2, t_4]$. Let $L_A$ and $L_B$ be the liquidity parameters of the positions of liquidity providers $A$ and $B$, respectively.

From Table 5-1, we obtain that

$$3 = L_A \left( \frac{1}{\sqrt{3,800}} - \frac{1}{\sqrt{t_3}} \right)$$

and hence, $L_A \approx 7,312.6996$.

In a similar way,

$$10 = L_B \left( \frac{1}{\sqrt{t_2}} - \frac{1}{\sqrt{t_4}} \right)$$

and thus, $L_B \approx 25,294.2194$.

Note that the initialized ticks are $t_1$, $t_2$, $t_3$, and $t_4$ and that the current price $p_0 = 3,800$ belongs to $[t_1, t_2]$. Note also that the liquidity of the pool in the interval $[t_1, t_2]$ is equal to $L_A$. We depict the situation in Figure 5-11.

**Trade 1.** Suppose that a trader wants to buy 1 ETH from the pool. By Equation 5.4, the virtual balances of the pool at the current price $p_0(3,800)$ are

$$x_v = \frac{L_A}{\sqrt{p_0}} \approx 118.62765 \quad \text{and} \quad y_v = L_A \sqrt{p_0} \approx 450,785.07922.$$

In addition, from Table 5-1, we obtain that the real balance of ETH at the current price (within the interval $[t_1, t_2]$) is

$$x_r = L_A \left( \frac{1}{\sqrt{3,800}} - \frac{1}{\sqrt{t_2}} \right) \approx 1.528.$$

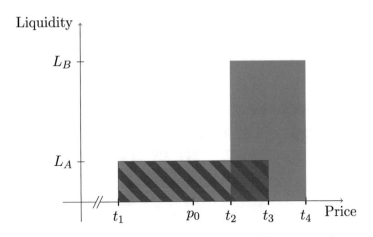

**Figure 5-11.** *Representation of the positions of the example*

Since the trader wants to buy 1 ETH and $x_r > 1$, it follows that the trade can be completed within the interval $[t_1, t_2]$.

From Equation 2.8, we obtain that the amount of USDC that the trader has to deposit in order to receive 1 ETH is

$$b_0 = \frac{1 \cdot y_v}{(1-\phi)(x_v - 1)} \approx 3,843.83684$$

After the trade, the spot price is updated as

$$p_1 = \left(\frac{L_A}{x_v - 1}\right)^2 \approx 3{,}864.8853.$$

Note that $3{,}864.8853 \in [t_1, t_2]$.

In addition, the trading fee of this trade is 0.3% of the amount that goes into the pool, and hence, it is approximately

$$3{,}843.83684 \cdot 0.003 \approx 11.53151 \text{ USDC}.$$

Thus,

$$f_g^1 = \frac{11.53151}{L_A} \approx 0.0015769.$$

Clearly, $f_g^0 = 0$ and also $f_o^j(i_k) = 0$ for all $j \in \{0, 1\}$ and for all $k \in \{1,2,3,4\}$.

**Trade 2.** Suppose now that another trader wants to deposit 5,000 USDC into the pool in order to buy ETH.

Following the analysis of Case 4 of the previous subsection, let $y_r$ be the real balance of USDC within the interval $[t_1, t_2]$ and let $y_{\max}$ be the maximum possible value of the real balance of USDC in the same interval. From Table 5-1, we know that

$$y_r = L_A\left(\sqrt{p_1} - \sqrt{t_1}\right) \approx 15{,}855.0899$$

and

$$y_{\max} = L_A\left(\sqrt{t_2} - \sqrt{t_1}\right) \approx 17{,}905.3157.$$

Since

$$y_r + 5{,}000 \cdot (1 - \phi) \approx 20{,}840.0899 > y_{max}$$

the trade cannot be completed within the interval $[t_1, t_2]$. Thus, a first step of the trade is performed, trading an amount $b_1'$ of USDC where

$$b_1' = y_{max} - y_r \approx 2{,}050.2258.$$

Note that this value of $b_1'$ is obtained from the equations of the Uniswap v3 protocol, where fees are not considered, since they are separated and kept aside. Hence, in order to have an amount $b_1'$ of USDC to trade (without fees being charged to it), the trader must provide an amount $b_1$ of USDC such that $0.997b_1 = b_1'$. Then, the fee that is charged to the trader in this first step is

$$0.003b_1 = \frac{0.003b_1'}{0.997} \approx 6.1692 \text{ USDC}.$$

Thus, the updated value of the variable $f_g^1$ is

$$f_g^1 \approx 0.0015769 + \frac{6.1692}{L_A} \approx 0.0024205$$

(see Table 5-4).

The amount $a_1$ of ETH that the trader receives for the first step of the trade can be computed using Equation 2.3 (since the fee is charged separately) with the updated virtual balances. We obtain that

$$a_1 = \frac{(x_v - 1)b_1'}{(y_v + b_0') + b_1'} \approx 0.528.$$

However, since the Uniswap v3 smart contract does not track the virtual balances, we will compute the amount $a_1$ in a different way. Observe that the virtual balances of the pool at the current price $p_1$ are

$$x_v = \frac{L_A}{\sqrt{p_1}} \approx 117.62765 \quad \text{and} \quad y_v = L_A\sqrt{p_1} \approx 454{,}617.3845.$$

After the first step of the trade, the updated virtual balance of USDC will be

$$y_v' = y_v + b_1' \approx 456{,}667.6103$$

and hence, the updated price will be

$$p' = \left(\frac{y_v'}{L_A}\right)^2 \approx 3{,}899.823492$$

that is, $p' = t_2$, as expected. The virtual balance of ETH at price $t_2$ (with respect to the interval $[t_1, t_2]$) is

$$x_v' = \frac{L_A}{\sqrt{t_2}} \approx 117.09956.$$

The difference between the virtual balances of ETH at prices $p_1$ and $t_2$ is the amount of ETH that the trader receives for the first step of the trade, that is,

$$a_1 = x_v - x_v' \approx 0.528.$$

Now, note that the amount to be traded is 5,000 USDC, which is greater than the amount $b_1$ used in the first step, since

$$b_1 = \frac{b_1'}{0.997} \approx 2{,}056.395.$$

Hence, tick $t_2$ needs to be crossed, and the remaining amount of USDC, which is $5{,}000 - b_1 \approx 2{,}943.605$, is going to be traded in the interval $[t_2, t_3]$. Before carrying out the second part of the trade, recall that the variables $L_{tot}, f_o^0(i_2)$, and $f_o^1(i_2)$ are updated since tick $t_2$ is crossed. Observe that the updated values of these variables are

$$\begin{aligned}
L_{tot} &= L_A + L_B \\
f_o^0(i_2) &= f_g^0 - f_o^0(i_2) = 0 - 0 = 0, \text{ and} \\
f_o^1(i_2) &= f_g^1 - f_o^1(i_2) = 0.0024205 - 0 = 0.0024205
\end{aligned}$$

(see Table 5-4).

Now we will perform the second part of the trade. Let $b_2$ be the remaining amount of USDC to be traded. As we pointed out before, $b_2 \approx 2{,}943.605$. Note that the current price is $t_2$ and that the real balance of USDC at price $t_2$ with respect to the interval $[t_2, t_3]$ is 0 (by the formulae of Table 5-1). Let $y_{max}'$ be the maximum possible value of the real balance of USDC in the interval $[t_2, t_3]$. From Table 5-1, we obtain that

$$y_{max}' = \left(L_A + L_B\right)\left(\sqrt{t_3} - \sqrt{t_2}\right) \approx 25{,}920.9386.$$

Let

$$b_2' = (1 - \phi)b_2 = 0.997 b_2 \approx 2{,}934.774.$$

Since $b_2'$ is clearly smaller than $y_{max}'$, the second step of the trade can be completed within the interval $[t_2, t_3]$. Note that $b_2'$ is the amount to be actually deposited in the pool and traded for ETH.

The fee that is charged in the second part of the trade is

$$0.003b_2 \approx 2{,}943.605 \cdot 0.003 \approx 8.8308 \text{ USDC}$$

and hence, the updated value of the variable $f_g^1$ is

$$f_g^1 \approx 0.0024205 + \frac{8.8308}{L_A + L_B} \approx 0.0026913.$$

Let $x_v$ and $y_v$ be the virtual balances of ETH and USDC, respectively, at the current price $t_2$ (in the interval $[t_2, t_3]$). By Equation 5.4,

$$x_v = \frac{L_A + L_B}{\sqrt{t_2}} \approx 522.1404$$

and

$$y_v = \left(L_A + L_B\right)\sqrt{t_2} \approx 2{,}036{,}255.3611.$$

After the trade, the updated virtual balance of USDC is

$$y_v' = y_v + b_2' \approx 2{,}039{,}190.1352,$$

and thus, the updated price is

$$p_3 = \left(\frac{y_v'}{L_A + L_B}\right)^2 \approx 3{,}911.0729.$$

Note that $3{,}911.0729 \in [t_2, t_3]$. Hence, the updated virtual balance of ETH is

$$x_v' = \frac{L_A + L_B}{\sqrt{p_3}} \approx 521.3889.$$

Let $a_2$ be the amount of ETH that the trader receives in the second step. Note that

$$a_2 = x_v - x_v' \approx 522.1404 - 521.3889 = 0.7515.$$

Therefore, the trader receives in total

$$a_1 + a_2 \approx 0.528 + 0.7515 = 1.2795 \text{ ETH}.$$

In Table 5-4, we summarize the price movements, the amounts of fees charged, and how the variables $f_g^1, f_o^1(i_2), f_a^1(i_2)$, and $f_b^1(i_2)$ are updated in the trades that were analyzed before. The values that are updated in each step are highlighted in bold type.

**Table 5-4.** *Summary of price movements and update of fee variables in trades 1 and 2*

| | Trade 1 | Trade 2 | | |
| | | Step 1 | Tick cross | Step 2 |
| --- | --- | --- | --- | --- |
| Price | $p_0 \rightarrow p_1$ | $p_1 \rightarrow t_2$ | $t_2$ | $t_2 \rightarrow p_3$ |
| Fee | 11.53151 | 6.1692 | 0 | 8.8308 |
| $f_g^1$ | **0.0015769** | **0.0024205** | 0.0024205 | **0.0026913** |
| $f_o^1(i_2)$ | 0 | 0 | **0.0024205** | 0.0024205 |
| $f_a^1(i_2)$ | 0 | 0 | 0 | **0.0002708** |
| $f_b^1(i_2)$ | **0.0015769** | **0.0024205** | 0.0024205 | 0.0024205 |

**Collected fees and impermanent loss.** We will now compute the amount of fees that each liquidity provider has obtained and find out whether they are facing an impermanent loss or not.

Note that the fees were all charged in USDC. Hence, we will only compute the amount of collected fees in this asset. The amount of fees that liquidity provider $A$ has collected is

$$
\begin{aligned}
L_A f_r^1(t_1, t_3) &= L_A\left(f_g^1 - f_b^1(i_1) - f_a^1(i_3)\right) \\
&= L_A\left(f_g^1 - f_o^1(i_1) - f_o^1(i_3)\right) \\
&= L_A\left(f_g^1 - 0 - 0\right) \approx 19.68,
\end{aligned}
$$

where the second equality follows from the fact that $t_1 < p_3 < t_3$.

On the other hand, the amount of fees that liquidity provider $B$ has collected is

$$
\begin{aligned}
L_B f_r^1(t_2, t_4) &= L_B\left(f_g^1 - f_b^1(i_2) - f_a^1(i_4)\right) \\
&= L_B\left(f_g^1 - f_o^1(i_2) - f_o^1(i_4)\right) \\
&= L_B\left(f_g^1 - f_o^1(i_2) - 0\right) \approx 6.85,
\end{aligned}
$$

where the second equality follows from the fact that $t_2 < p_3 < t_4$.

We will now analyze the existence of impermanent losses. Recall that liquidity provider $A$ deposited 3 ETH and a certain amount of USDC when the price was 3,800 and the liquidity parameter of their position was $L_A \approx 7{,}312.6996$. From Table 5-1, we obtain that the amount of USDC that they had to deposit was

$$
L_A\left(\sqrt{3{,}800} - \sqrt{t_1}\right) \approx 12{,}022.78454.
$$

Hence, if they had never deposited their assets into the pool, the value of their position at the current price $p_3$ would have been

$$V_A \approx 3 \cdot p_3 + 12{,}022.78454 \approx 23{,}756 \text{ USDC.}$$

On the other hand, applying the formulae of Table 5-1, we obtain that the real balances of their position (at the current price $p_3$) are

$$L_A\left(\frac{1}{\sqrt{p_3}} - \frac{1}{\sqrt{t_3}}\right) \approx 1.303378 \text{ ETH.}$$

and

$$L_A\left(\sqrt{p_3} - \sqrt{t_1}\right) \approx 18{,}563.49259 \text{ USDC.}$$

Hence, the current value of their position is

$$V_A' \approx 1.303378 p_3 + 18{,}563.49259 \approx 23{,}661.1 \text{ USDC.}$$

Therefore, liquidity provider $A$ is facing an impermanent loss of approximately $23{,}756 - 23{,}661.1 = 94.9$ USDC, which is not compensated by the fees of 19.68 USDC they have earned.

On the other hand, liquidity provider $B$ deposited 10 ETH into the pool (and 0 USDC, since the price at the moment of the deposit was below the interval they chose). If they had never deposited that amount of ETH into the pool, the value of their position at the current price of $p_3$ would have been

$$V_B \approx 10 \cdot p_3 \approx 39{,}110.73 \text{ USDC.}$$

From Table 5-1, we get that at price $p_3$, the real balances of their position are

$$L_B \left( \frac{1}{\sqrt{p_3}} - \frac{1}{\sqrt{t_4}} \right) \approx 9.4170708 \text{ ETH.}$$

and

$$L_B \left( \sqrt{p_3} - \sqrt{t_2} \right) \approx 2{,}276.59725 \text{ USDC.}$$

Hence, the current value of their position is

$$V_B^{'} \approx 9.4170708 p_3 + 2{,}276.59725 \approx 39{,}107.45 \text{ USDC.}$$

Therefore, liquidity provider $B$ is facing an impermanent loss of approximately $39{,}110.73 - 39{,}107.45 = 3.28$ USDC, which, in this case, is compensated by the fees of 6.85 USDC they have earned. Observe that the impermanent loss for liquidity provider $B$ is very small because the amounts of ETH and USDC that their position currently has are very similar to those of their original deposit. This is so because although they entered the position at a price of 3,800, they provided liquidity in the interval $[t_2, t_4]$ (which is approximately [3,900, 4,100]), and hence, their position did not change when the price moved from 3,800 to 3,900 (neither did they earn any trading fees). Note that the current price is $p_3 \approx 3{,}911.0729$, which is very near $t_2 \approx 3{,}900$.

## 5.6  LP Tokens

Unlike Uniswap v2, LP tokens in Uniswap v3 are nonfungible tokens (NFTs) since positions are highly customizable and will be, in general, very different for different liquidity providers. When a liquidity provider adds liquidity to a specific price range, a unique NFT defining the position is minted. Recall that in Uniswap v3, a liquidity provider's position consists of the address of the owner, the lower tick index, and the upper tick index. These last two define the boundaries of the interval in which the liquidity provider deposited liquidity. The owner of the NFT that represents a certain position can modify that position or redeem the corresponding tokens.

Recall also that each position stores several variables, as we can see in Listing 5-11.[4]

***Listing 5-11.*** Position library of the smart contract of Uniswap v3

```
/// @title Position
/// @notice Positions represent an owner address'
  ↪liquidity between a lower and upper tick boundary
/// @dev Positions store additional state for tracking
  ↪fees owed to the position
library Position {
    // info stored for each user's position
    struct Info {
        // the amount of liquidity owned by this
  ↪position
        uint128 liquidity;
        // fee growth per unit of liquidity as of the
  ↪last update to liquidity or fees owed
        uint256 feeGrowthInside0LastX128;
        uint256 feeGrowthInside1LastX128;
        // the fees owed to the position owner in
  ↪token0/token1
        uint128 tokensOwed0;
        uint128 tokensOwed1;
    }
```

---

[4] This code can be found in `https://github.com/Uniswap/v3-core/blob/main/contracts/libraries/Position.sol`

```
    /// @notice Returns the Info struct of a position,
↪given an owner and position boundaries
    /// @param self The mapping containing all user
↪positions
    /// @param owner The address of the position owner
    /// @param tickLower The lower tick boundary of the
↪position
    /// @param tickUpper The upper tick boundary of the
↪position
    /// @return position The position info struct of
↪the given owners' position
    function get(
        mapping(bytes32 => Info) storage self,
        address owner,
        int24 tickLower,
        int24 tickUpper
    ) internal view returns (Position.Info storage
↪position) {
        position =
↪self[keccak256(abi.encodePacked(owner, tickLower,
↪tickUpper))];
    }

    /// @notice Credits accumulated fees to a user's
↪position
    /// @param self The individual position to update
    /// @param liquidityDelta The change in pool
↪liquidity as a result of the position update
    /// @param feeGrowthInsideOX128 The all-time fee
↪growth in token0, per unit of liquidity, inside the
↪position's tick boundaries
```

```
/// @param feeGrowthInside1X128 The all-time fee
↪growth in token1, per unit of liquidity, inside the
↪position's tick boundaries
  function update(
      Info storage self,
      int128 liquidityDelta,
      uint256 feeGrowthInside0X128,
      uint256 feeGrowthInside1X128
  ) internal {
      Info memory _self = self;

      uint128 liquidityNext;
      if (liquidityDelta == 0) {
          require(_self.liquidity > 0, 'NP'); //
↪disallow pokes for 0 liquidity positions
          liquidityNext = _self.liquidity;
      } else {
          liquidityNext =
↪LiquidityMath.addDelta(_self.liquidity,
↪liquidityDelta);
      }

      // calculate accumulated fees
      uint128 tokensOwed0 =
          uint128(
              FullMath.mulDiv(
                  feeGrowthInside0X128 -
↪_self.feeGrowthInside0LastX128,
                  _self.liquidity,
                  FixedPoint128.Q128
              )
          );
```

```
            uint128 tokensOwed1 =
                uint128(
                    FullMath.mulDiv(
                        feeGrowthInside1X128 -
↪_self.feeGrowthInside1LastX128,
                        _self.liquidity,
                        FixedPoint128.Q128
                    )
                );

        // update the position
        if (liquidityDelta != 0) self.liquidity =
↪liquidityNext;
        self.feeGrowthInside0LastX128 =
↪feeGrowthInside0X128;
        self.feeGrowthInside1LastX128 =
↪feeGrowthInside1X128;
        if (tokensOwed0 > 0 || tokensOwed1 > 0) {
            // overflow is acceptable, have to withdraw
↪before you hit type(uint128).max fees
            self.tokensOwed0 += tokensOwed0;
            self.tokensOwed1 += tokensOwed1;
        }
    }
}
```

We also point out that the Position class has two methods: get and update, as we could see in the previous code. The get method returns the Info struct of a position, while the update method credits the accumulated fees to a position and, if necessary, updates the liquidity parameter of that position.

# 5.6.1 Minting LP Tokens

The input requirements when adding liquidity to the pool–that is, when LP tokens are going to be minted–are the *owner*, the *liquidity* added, and the price range defined by the *lower* and *upper* tick indexes. The function that is called to perform a new liquidity deposit is the mint function given in Listing 5-12.[5]

***Listing 5-12.*** mint function of the smart contract of Uniswap v3

```
/// @inheritdoc IUniswapV3PoolActions
/// @dev noDelegateCall is applied indirectly via
↪_modifyPosition
  function mint(
      address recipient,
      int24 tickLower,
      int24 tickUpper,
      uint128 amount,
      bytes calldata data
  ) external override lock returns (uint256 amount0,
↪uint256 amount1) {
      require(amount > 0);
      (, int256 amount0Int, int256 amount1Int) =
          _modifyPosition(
              ModifyPositionParams({
                  owner: recipient,
                  tickLower: tickLower,
                  tickUpper: tickUpper,
                  liquidityDelta:
```

---

[5] This code can be found in https://github.com/Uniswap/v3-core/blob/main/contracts/UniswapV3Pool.sol

```
↪int256(amount).toInt128()
            })
        );

    amount0 = uint256(amount0Int);
    amount1 = uint256(amount1Int);

    uint256 balance0Before;
    uint256 balance1Before;
    if (amount0 > 0) balance0Before = balance0();
    if (amount1 > 0) balance1Before = balance1();

↪IUniswapV3MintCallback(msg.sender).uniswapV3MintCallback(amount0
↪amount1, data);
    if (amount0 > 0)
↪require(balance0Before.add(amount0) <= balance0(),
↪'M0');
    if (amount1 > 0)
↪require(balance1Before.add(amount1) <= balance1(),
↪'M1');

    emit Mint(msg.sender, recipient, tickLower,
↪tickUpper, amount, amount0, amount1);
  }
```

## 5.6.2  Modifying the Position: Part I

One of the most important parts of the mint function is the function
_modifyPosition, which creates a new position and allows the owner to
modify it. The parameters that are needed in order to modify the position
are the owner, the tick range, and the liquidity that is going to be added or
removed. This is shown in Listing 5-13.

***Listing 5-13.*** Parameters needed to modify a position

```
struct ModifyPositionParams {
    // the address that owns the position
    address owner;
    // the lower and upper tick of the position
    int24 tickLower;
    int24 tickUpper;
    // any change in liquidity
    int128 liquidityDelta;
}
```

The position is updated using the previous parameters, as we can see in Listing 5-14, where the _updatePosition function is called.

***Listing 5-14.*** First part of the _modifyPosition function of the smart contract of Uniswap v3

```
/// @dev Effect some changes to a position
/// @param params the position details and the
↪change to the position's liquidity to effect
/// @return position a storage pointer referencing
↪the position with the given owner and tick range
/// @return amount0 the amount of token0 owed to
↪the pool, negative if the pool should pay the
↪recipient
/// @return amount1 the amount of token1 owed to
↪the pool, negative if the pool should pay the
↪recipient
function _modifyPosition(ModifyPositionParams
↪memory params)
    private
    noDelegateCall
```

```
    returns (
        Position.Info storage position,
        int256 amount0,
        int256 amount1
    )
{

    checkTicks(params.tickLower, params.tickUpper);

    Slot0 memory _slot0 = slot0; // SLOAD for gas
↪optimization

    position = _updatePosition(
        params.owner,
        params.tickLower,
        params.tickUpper,
        params.liquidityDelta,
        _slot0.tick
    );
    ...
```

## 5.6.3 Update the Position

The function _updatePosition, shown in Listing 5-15, updates the
liquidity of the position and computes and stores the accumulated fees
since the last update, as we can see in the following code. It also initializes
or uninitializes the boundary tick indexes of the position in case either of
these actions is needed (e.g., if the position is created or removed and the
boundary tick indexes are not referenced by any other position).

***Listing 5-15.*** _updatePosition function of the smart contract of Uniswap v3

```
/// @dev Gets and updates a position with the given
↪liquidity delta
/// @param owner the owner of the position
/// @param tickLower the lower tick of the
↪position's tick range
/// @param tickUpper the upper tick of the
↪position's tick range
/// @param tick the current tick, passed to avoid
↪sloads
  function _updatePosition(
      address owner,
      int24 tickLower,
      int24 tickUpper,
      int128 liquidityDelta,
      int24 tick
  ) private returns (Position.Info storage position)
↪{
      position = positions.get(owner, tickLower,
↪tickUpper);
      uint256 _feeGrowthGlobal0X128 =
↪feeGrowthGlobal0X128; // SLOAD for gas
↪optimization
      uint256 _feeGrowthGlobal1X128 =
↪feeGrowthGlobal1X128; // SLOAD for gas
↪optimization

      // if we need to update the ticks, do it
      bool flippedLower;
      bool flippedUpper;
      if (liquidityDelta != 0) {
```

```
    uint32 time = _blockTimestamp();
    (int56 tickCumulative, uint160
↪secondsPerLiquidityCumulativeX128) =
        observations.observeSingle(
            time,
            0,
            slot0.tick,
            slot0.observationIndex,
            liquidity,
            slot0.observationCardinality
        );

flippedLower = ticks.update(
    tickLower,
    tick,
    liquidityDelta,
    _feeGrowthGlobal0X128,
    _feeGrowthGlobal1X128,
    secondsPerLiquidityCumulativeX128,
    tickCumulative,
    time,
    false,
    maxLiquidityPerTick
);
flippedUpper = ticks.update(
    tickUpper,
    tick,
    liquidityDelta,
    _feeGrowthGlobal0X128,
    _feeGrowthGlobal1X128,
    secondsPerLiquidityCumulativeX128,
    tickCumulative,
```

```
                        time,
                        true,
                        maxLiquidityPerTick
                );

                if (flippedLower) {
                        tickBitmap.flipTick(tickLower, tickSpacing);
                }
                if (flippedUpper) {
                        tickBitmap.flipTick(tickUpper, tickSpacing);
                }
        }

        (uint256 feeGrowthInside0X128, uint256
↪feeGrowthInside1X128) =
                ticks.getFeeGrowthInside(tickLower,
↪tickUpper, tick, _feeGrowthGlobal0X128,
↪ _feeGrowthGlobal1X128);

        position.update(liquidityDelta,
↪feeGrowthInside0X128, feeGrowthInside1X128);

        // clear any tick data that is no longer
↪needed
        if (liquidityDelta < 0) {
                if (flippedLower) {
                        ticks.clear(tickLower);
                }
                if (flippedUpper) {
                        ticks.clear(tickUpper);
                }
        }
    }
```

## 5.6.4 Tick Class

In the previous code, the Tick class[6] appeared. Its code is given in Listing 5-16. Observe that the Tick class keeps track of the following variables:

> liquidityGross
>
> liquidityNet
>
> feeGrowthOutside0X128
>
> feeGrowthOutside1X128

which, in the previous section, were denoted by $L_g$, $\Delta L$, $f_o^0$, and $f_o^1$, respectively. Note also that the method getFeeGrowthInside defines the following variables:

> feeGrowthBelow0X128
>
> feeGrowthBelow 1X128
>
> feegrowthAbove0X128
>
> feegrowthAbove 1X128
>
> feeGrowthInside0X128
>
> feeGrowthInside1X128

which correspond to the variables $f_b^0$, $f_b^1$, $f_a^0$, $f_a^1$, $f_r^0$, and $f_r^1$, respectively, that were defined in the previous section.

***Listing 5-16.*** Tick class of the smart contract of Uniswap v3

```
/// @title Tick
/// @notice Contains functions for managing tick
  ↪processes and relevant calculations
```

---

[6] The Tick class can be found in https://github.com/Uniswap/v3-core/blob/main/contracts/libraries/Tick.sol

```
library Tick {
    using LowGasSafeMath for int256;
    using SafeCast for int256;

    // info stored for each initialized individual
    ↪tick
      struct Info {
        // the total position liquidity that references
        ↪this tick
        uint128 liquidityGross;
        // amount of net liquidity added (subtracted)
        ↪when tick is crossed from left to right (right to
        ↪left),
        int128 liquidityNet;
        // fee growth per unit of liquidity on the
  ↪_other_ side of this tick (relative to the current
  ↪tick)
        // only has relative meaning, not absolute -
  ↪the value depends on when the tick is initialized
        uint256 feeGrowthOutside0X128;
        uint256 feeGrowthOutside1X128;
        // the cumulative tick value on the other side
  ↪of the tick
        int56 tickCumulativeOutside;
        // the seconds per unit of liquidity on the
  ↪_other_ side of this tick (relative to the current
  ↪tick)
        // only has relative meaning, not absolute -
  ↪the value depends on when the tick is initialized
        uint160 secondsPerLiquidityOutsideX128;
        // the seconds spent on the other side of the
  ↪tick (relative to the current tick)
```

```
    // only has relative meaning, not absolute -
↪the value depends on when the tick is initialized
    uint32 secondsOutside;
    // true iff the tick is initialized, i.e. the
↪value is exactly equivalent to the expression
↪liquidityGross != 0
    // these 8 bits are set to prevent fresh
↪sstores when crossing newly initialized ticks
    bool initialized;
  }

  /// @notice Derives max liquidity per tick from
↪given tick spacing
  /// @dev Executed within the pool constructor
  /// @param tickSpacing The amount of required tick
↪separation, realized in multiples of `tickSpacing`
  ///     e.g., a tickSpacing of 3 requires ticks to
↪be initialized every 3rd tick i.e., ..., -6, -3, 0,
↪3, 6, ...
  /// @return The max liquidity per tick
  function tickSpacingToMaxLiquidityPerTick(int24
↪tickSpacing) internal pure returns (uint128) {
    int24 minTick = (TickMath.MIN_TICK /
↪tickSpacing) * tickSpacing;
    int24 maxTick = (TickMath.MAX_TICK /
↪tickSpacing) * tickSpacing;
    uint24 numTicks = uint24((maxTick - minTick) /
↪tickSpacing) + 1;
    return type(uint128).max / numTicks;
  }
```

```
/// @notice Retrieves fee growth data
/// @param self The mapping containing all tick
↪information for initialized ticks
/// @param tickLower The lower tick boundary of the
↪position
/// @param tickUpper The upper tick boundary of the
↪position
/// @param tickCurrent The current tick
/// @param feeGrowthGlobal0X128 The all-time global
↪fee growth, per unit of liquidity, in token0
/// @param feeGrowthGlobal1X128 The all-time global
↪fee growth, per unit of liquidity, in token1
/// @return feeGrowthInside0X128 The all-time fee
↪growth in token0, per unit of liquidity, inside the
↪position's tick boundaries
/// @return feeGrowthInside1X128 The all-time fee
↪growth in token1, per unit of liquidity, inside the
↪position's tick boundaries
function getFeeGrowthInside(
    mapping(int24 => Tick.Info) storage self,
    int24 tickLower,
    int24 tickUpper,
    int24 tickCurrent,
    uint256 feeGrowthGlobal0X128,
    uint256 feeGrowthGlobal1X128
) internal view returns (uint256
↪feeGrowthInside0X128, uint256 feeGrowthInside1X128)
↪{
    Info storage lower = self[tickLower];
    Info storage upper = self[tickUpper];
```

```
    // calculate fee growth below
    uint256 feeGrowthBelow0X128;
    uint256 feeGrowthBelow1X128;
    if (tickCurrent >= tickLower) {
        feeGrowthBelow0X128 =
↪lower.feeGrowthOutside0X128;
        feeGrowthBelow1X128 =
↪lower.feeGrowthOutside1X128;
    } else {
        feeGrowthBelow0X128 = feeGrowthGlobal0X128
↪- lower.feeGrowthOutside0X128;
        feeGrowthBelow1X128 = feeGrowthGlobal1X128
↪- lower.feeGrowthOutside1X128;
    }

    // calculate fee growth above
    uint256 feeGrowthAbove0X128;
    uint256 feeGrowthAbove1X128;
    if (tickCurrent < tickUpper) {
        feeGrowthAbove0X128 =
↪upper.feeGrowthOutside0X128;
        feeGrowthAbove1X128 =
↪upper.feeGrowthOutside1X128;
    } else {
        feeGrowthAbove0X128 = feeGrowthGlobal0X128
↪- upper.feeGrowthOutside0X128;
        feeGrowthAbove1X128 = feeGrowthGlobal1X128
↪- upper.feeGrowthOutside1X128;
    }

    feeGrowthInside0X128 = feeGrowthGlobal0X128 -
↪feeGrowthBelow0X128 - feeGrowthAbove0X128;
```

```
        feeGrowthInside1X128 = feeGrowthGlobal1X128 -
↪feeGrowthBelow1X128 - feeGrowthAbove1X128;
    }
```

```
    /// @notice Updates a tick and returns true if the
↪tick was flipped from initialized to uninitialized,
↪or vice versa
    /// @param self The mapping containing all tick
↪information for initialized ticks
    /// @param tick The tick that will be updated
    /// @param tickCurrent The current tick
    /// @param liquidityDelta A new amount of liquidity
↪to be added (subtracted) when tick is crossed from
↪left to right (right to left)
    /// @param feeGrowthGlobal0X128 The all-time global
↪fee growth, per unit of liquidity, in token0
    /// @param feeGrowthGlobal1X128 The all-time global
↪fee growth, per unit of liquidity, in token1
    /// @param secondsPerLiquidityCumulativeX128 The
↪all-time seconds per max(1, liquidity) of the pool
    /// @param tickCumulative The tick * time elapsed
↪since the pool was first initialized
    /// @param time The current block timestamp cast to
↪a uint32
    /// @param upper true for updating a position's
↪upper tick, or false for updating a position's
↪lower tick
    /// @param maxLiquidity The maximum liquidity
↪allocation for a single tick
    /// @return flipped Whether the tick was flipped
↪from initialized to uninitialized, or vice versa
```

```
function update(
    mapping(int24 => Tick.Info) storage self,
    int24 tick,
    int24 tickCurrent,
    int128 liquidityDelta,
    uint256 feeGrowthGlobal0X128,
    uint256 feeGrowthGlobal1X128,
    uint160 secondsPerLiquidityCumulativeX128,
    int56 tickCumulative,
    uint32 time,
    bool upper,
    uint128 maxLiquidity
) internal returns (bool flipped) {
    Tick.Info storage info = self[tick];

    uint128 liquidityGrossBefore =
↪info.liquidityGross;
    uint128 liquidityGrossAfter =
↪LiquidityMath.addDelta(liquidityGrossBefore,
↪liquidityDelta);

    require(liquidityGrossAfter <= maxLiquidity,
↪'LO');

    flipped = (liquidityGrossAfter == 0) !=
↪(liquidityGrossBefore == 0);

    if (liquidityGrossBefore == 0) {
        // by convention, we assume that all growth
↪before a tick was initialized happened _below_ the
↪tick
```

```
        if (tick <= tickCurrent) {
            info.feeGrowthOutside0X128 =
↪feeGrowthGlobal0X128;
            info.feeGrowthOutside1X128 =
↪feeGrowthGlobal1X128;
            info.secondsPerLiquidityOutsideX128 =
↪secondsPerLiquidityCumulativeX128;
            info.tickCumulativeOutside =
↪tickCumulative;
            info.secondsOutside = time;
        }
        info.initialized = true;
    }

    info.liquidityGross = liquidityGrossAfter;

    // when the lower (upper) tick is crossed left
↪to right (right to left), liquidity must be added
↪(removed)
    info.liquidityNet = upper
        ?
↪int256(info.liquidityNet).sub(liquidityDelta).toInt128()
        :
↪int256(info.liquidityNet).add(liquidityDelta).toInt128();
    }

    /// @notice Clears tick data
    /// @param self The mapping containing all
↪initialized tick information for initialized ticks
    /// @param tick The tick that will be cleared
    function clear(mapping(int24 => Tick.Info) storage
↪self, int24 tick) internal {
```

```
        delete self[tick];
    }

    /// @notice Transitions to next tick as needed by
↪price movement
    /// @param self The mapping containing all tick
↪information for initialized ticks
    /// @param tick The destination tick of the
↪transition
    /// @param feeGrowthGlobal0X128 The all-time global
↪fee growth, per unit of liquidity, in token0
    /// @param feeGrowthGlobal1X128 The all-time global
↪fee growth, per unit of liquidity, in token1
    /// @param secondsPerLiquidityCumulativeX128 The
↪current seconds per liquidity
    /// @param tickCumulative The tick * time elapsed
↪since the pool was first initialized
    /// @param time The current block.timestamp
    /// @return liquidityNet The amount of liquidity
↪added (subtracted) when tick is crossed from left
↪to right (right to left)
    function cross(
        mapping(int24 => Tick.Info) storage self,
        int24 tick,
        uint256 feeGrowthGlobal0X128,
        uint256 feeGrowthGlobal1X128,
        uint160 secondsPerLiquidityCumulativeX128,
        int56 tickCumulative,
        uint32 time
```

```
) internal returns (int128 liquidityNet) {
    Tick.Info storage info = self[tick];
    info.feeGrowthOutside0X128 =
↪feeGrowthGlobal0X128 - info.feeGrowthOutside0X128;
    info.feeGrowthOutside1X128 =
↪feeGrowthGlobal1X128 - info.feeGrowthOutside1X128;
    info.secondsPerLiquidityOutsideX128 =
↪secondsPerLiquidityCumulativeX128 -
↪info.secondsPerLiquidityOutsideX128;
    info.tickCumulativeOutside = tickCumulative -
↪info.tickCumulativeOutside;
    info.secondsOutside = time -
↪info.secondsOutside;
    liquidityNet = info.liquidityNet;
}
}
```

### Modifying the Position: Part II

The second part of the _modifyPosition function computes the amounts of each of the pool tokens that the liquidity provider needs to deposit in order to set up the position. These amounts are called amount0 and amount1 and are returned by the _modifyPosition function. Recall that the formulae for computing these amounts are given in Table 5-1. We show the whole _modifyPosition function in Listing 5-17.

*Listing 5-17.* _modifyPosition function of the smart contract of Uniswap v3

```
/// @dev Effect some changes to a position
/// @param params the position details and the
↪change to the position's liquidity to effect
/// @return position a storage pointer referencing
↪the position with the given owner and tick range
```

```
/// @return amount0 the amount of token0 owed to
↪the pool, negative if the pool should pay the
↪recipient
/// @return amount1 the amount of token1 owed to
↪the pool, negative if the pool should pay the
↪recipient
function _modifyPosition(ModifyPositionParams
↪memory params)
    private
    noDelegateCall
    returns (
        Position.Info storage position,
        int256 amount0,
        int256 amount1
    )
{
    checkTicks(params.tickLower, params.tickUpper);

    Slot0 memory _slot0 = slot0; // SLOAD for gas
↪optimization

    position = _updatePosition(
        params.owner,
        params.tickLower,
        params.tickUpper,
        params.liquidityDelta,
        _slot0.tick
    );

    if (params.liquidityDelta != 0) {
        if (_slot0.tick < params.tickLower) {
            // current tick is below the passed
↪range; liquidity can only become in range by
```

```
↪crossing from left to
                // right, when we'll need _more_ token0
↪(it's becoming more valuable) so user must provide
↪it
                amount0 =
↪SqrtPriceMath.getAmount0Delta(

↪TickMath.getSqrtRatioAtTick(params.tickLower),

↪TickMath.getSqrtRatioAtTick(params.tickUpper),
                    params.liquidityDelta
                );
            } else if (_slot0.tick < params.tickUpper)
↪{
                // current tick is inside the passed
↪range
                uint128 liquidityBefore = liquidity; //
↪SLOAD for gas optimization
                // write an oracle entry
                (slot0.observationIndex,
↪slot0.observationCardinality) = observations.write(
                    _slot0.observationIndex,
                    _blockTimestamp(),
                    _slot0.tick,
                    liquidityBefore,
                    _slot0.observationCardinality,
                    _slot0.observationCardinalityNext
                );

                amount0 =
↪SqrtPriceMath.getAmount0Delta(
                    _slot0.sqrtPriceX96,
```

```
↪TickMath.getSqrtRatioAtTick(params.tickUpper),
                params.liquidityDelta
            );
            amount1 =
↪SqrtPriceMath.getAmount1Delta(

↪TickMath.getSqrtRatioAtTick(params.tickLower),
                _slot0.sqrtPriceX96,
                params.liquidityDelta
            );

            liquidity =
↪LiquidityMath.addDelta(liquidityBefore,
↪params.liquidityDelta);
        } else {
            // current tick is above the passed
↪range; liquidity can only become in range by
↪crossing from right to
            // left, when we'll need _more_ token1
  ↪(it's becoming more valuable) so user must provide
  ↪it
            amount1 =
↪SqrtPriceMath.getAmount1Delta(

↪TickMath.getSqrtRatioAtTick(params.tickLower),

↪TickMath.getSqrtRatioAtTick(params.tickUpper),
                params.liquidityDelta
            );
        }
```

Observe that the previous code uses the functions getAmount0Delta and getAmount1Delta from the SqrtPriceMath library.[7] These functions are given in Listing 5-18.

***Listing 5-18.*** Functions getAmount0Delta and getAmount1Delta of the smart contract of Uniswap v3

```
/// @notice Gets the amount0 delta between two
↪prices
/// @dev Calculates liquidity / sqrt(lower) -
↪liquidity / sqrt(upper),
/// i.e. liquidity * (sqrt(upper) - sqrt(lower)) /
↪(sqrt(upper) * sqrt(lower))
/// @param sqrtRatioAX96 A sqrt price
/// @param sqrtRatioBX96 Another sqrt price
/// @param liquidity The amount of usable
↪liquidity
/// @param roundUp Whether to round the amount up
↪or down
/// @return amount0 Amount of token0 required to
↪cover a position of size liquidity between the two
↪passed prices
function getAmount0Delta(
    uint160 sqrtRatioAX96,
    uint160 sqrtRatioBX96,
    uint128 liquidity,
    bool roundUp
) internal pure returns (uint256 amount0) {
    if (sqrtRatioAX96 > sqrtRatioBX96)
```

---

[7] https://github.com/Uniswap/v3-core/blob/main/contracts/libraries/ SqrtPriceMath.sol

```
↪(sqrtRatioAX96, sqrtRatioBX96) = (sqrtRatioBX96,
↪sqrtRatioAX96);
      uint256 numerator1 = uint256(liquidity) <<
↪FixedPoint96.RESOLUTION;
      uint256 numerator2 = sqrtRatioBX96 -
↪sqrtRatioAX96;

      require(sqrtRatioAX96 > 0);

      return
          roundUp
              ? UnsafeMath.divRoundingUp(
↪FullMath.mulDivRoundingUp(numerator1, numerator2,
↪sqrtRatioBX96),
                    sqrtRatioAX96
              )
              : FullMath.mulDiv(numerator1,
↪numerator2, sqrtRatioBX96) / sqrtRatioAX96;
  }
  /// @notice Gets the amount1 delta between two
↪prices
  /// @dev Calculates liquidity * (sqrt(upper) -
↪sqrt(lower))
  /// @param sqrtRatioAX96 A sqrt price
  /// @param sqrtRatioBX96 Another sqrt price
  /// @param liquidity The amount of usable
↪liquidity
  /// @param roundUp Whether to round the amount up,
↪or down
  /// @return amount1 Amount of token1 required to
↪cover a position of size liquidity between the two
↪passed prices
```

```
    function getAmount1Delta(
        uint160 sqrtRatioAX96,
        uint160 sqrtRatioBX96,
        uint128 liquidity,
        bool roundUp
    ) internal pure returns (uint256 amount1) {
        if (sqrtRatioAX96 > sqrtRatioBX96)
↪(sqrtRatioAX96, sqrtRatioBX96) = (sqrtRatioBX96,
↪sqrtRatioAX96);

        return
            roundUp
                ? FullMath.mulDivRoundingUp(liquidity,
↪sqrtRatioBX96 - sqrtRatioAX96, FixedPoint96.Q96)
                : FullMath.mulDiv(liquidity,
↪sqrtRatioBX96 - sqrtRatioAX96, FixedPoint96.Q96);
    }
```

Observe also that in the code of the _modifyPosition function, the amounts

```
slot0.sqrtPriceX96 ,
TickMath.getSqrtRatioAtTick(params.tickLower) , and
TickMath.getSqrtRatioAtTick(params.tickUpper)
```

correspond to the variables $\sqrt{p}, \sqrt{p_a}$, and $\sqrt{p_b}$ , respectively, that are used in Table 5-1.

## 5.6.5  Burning LP Tokens

The owner of LP tokens of a Uniswap v3 pool can remove liquidity. The liquidity removal process will burn the LP tokens, remove the position, and give back to the liquidity provider a certain amount of assets, according to the formulae of Table 5-1. In addition, the fees that accumulated since the last time they were collected are given to the liquidity provider.

As we can see in Listing 5-19, the burn function makes use of the _modifyPosition function that we studied before. In this case, a negative value for the liquidity is used, since liquidity is removed instead of being deposited.

***Listing 5-19.*** burn function of the smart contract of Uniswap v3

```
/// @inheritdoc IUniswapV3PoolActions
/// @dev noDelegateCall is applied indirectly via
↪_modifyPosition
  function burn(
      int24 tickLower,
      int24 tickUpper,
      uint128 amount
  ) external override lock returns (uint256 amount0,
↪uint256 amount1) {
      (Position.Info storage position, int256
↪amount0Int, int256 amount1Int) =
          _modifyPosition(
              ModifyPositionParams({
                  owner: msg.sender,
                  tickLower: tickLower,
                  tickUpper: tickUpper,
                  liquidityDelta:
↪- int256(amount).toInt128()
              })
          );

      amount0 = uint256(-amount0Int);
      amount1 = uint256(-amount1Int);
```

```
    if (amount0 > 0 || amount1 > 0) {
        (position.tokensOwed0,
↪position.tokensOwed1) = (
            position.tokensOwed0 +
↪uint128(amount0),
            position.tokensOwed1 + uint128(amount1)
        );
    }

    emit Burn(msg.sender, tickLower, tickUpper,
↪amount, amount0, amount1);
  }
```

# 5.7  Analysis of Liquidity Provisioning

Compared to Uniswap v2, in Uniswap v3, liquidity is provided with higher capital efficiency, meaning that with the same amount of deposited tokens, one can obtain a higher liquidity parameter in the chosen interval. This is beneficial for traders, since the higher the liquidity parameter is, the lower the price impact will be. However, for liquidity providers, the situation is very different. In order to analyze this, we will consider two situations with two different assumptions.

First, suppose that the exact same trades are executed in a Uniswap v2 pool and in a Uniswap v3 pool. In this case, the amount of trading fees collected by both pools will be the same–for example, 0.3% of the incoming amount of each trade–and the total collected fees will be shared between all the liquidity providers. In the Uniswap v2 pool, the fees will be shared in a way that is proportional to the deposits the liquidity providers made, while in the Uniswap v3 pool, this proportion will also depend on the interval that each liquidity provider chose. More explicitly, in the case of the Uniswap v3 pool, as we have seen, the liquidity providers earn fees only for the trades (or portion of trades) that are performed within the interval they chose, and in this case, their share of fees is equal to the liquidity parameter they provided divided by the total liquidity–which will also be higher than in a Uniswap v2 pool. To sum up, comparing a Uniswap v2 pool with a Uniswap v3 pool, if the same trades are performed, the total collected fees will be the same, and thus, liquidity providers will not be able to obtain more fees in general with the Uniswap v3 pool, although it may happen that some of them earn more fees with respect to a Uniswap v2 pool and some of them earn fewer fees.

For the second situation, we will consider a completely different assumption. One can argue that a considerable part of the trades that are executed in a Uniswap pool (either v2 or v3) is done by arbitrageurs and that arbitrageurs have enough money to perform any required trade. Thus, every time the spot price of the pool deviates from the market price, arbitrageurs will step in and make a trade. Due to the concentrated liquidity of Uniswap v3, in the price range that is near the market price, we usually have far more liquidity than in a Uniswap v2 pool. Hence, the arbitrageurs of Uniswap v3 will need to perform a bigger trade to make the spot price equal to the market price, or in other words, they will have the opportunity to make more money with the arbitrage. Since a bigger trade is performed, a larger amount of fees is collected, which may, at first sight, seem beneficial to liquidity providers. However, as we have seen in Section 5.3, concentrated liquidity also implies a higher impermanent loss for liquidity providers in many different possible scenarios. In addition, several analyses show that the possible greater amount of fees that liquidity providers collect (under the assumptions of this paragraph) does not compensate for the higher impermanent losses [25, 33].

## 5.7.1  Capital Efficiency

In order to compute the amount of liquidity that a liquidity provider puts into a Uniswap v3 pool, we will assume that the entry price is the geometric mean of the boundaries of the interval in which they provide liquidity. Hence, if the current price is $p$, we will assume that the liquidity provider chooses a price interval $[p_a, p_b]$ such that $\sqrt{p_a p_b} = p$. With this assumption, the amounts of each token that are needed to provide liquidity in a Uniswap v3 pool will satisfy the deposit requirements of a Uniswap v2 pool, as we shall see in the following text.

But first note that

$$p = p_a \sqrt{\frac{p_b}{p_a}} \quad \text{and} \quad p_b = p_a \left( \sqrt{\frac{p_b}{p_a}} \right)^2.$$

This means that if, for example, $p$ is 20% higher than $p_a$, then $p_b$ is 20% higher than $p$.

Consider a Uniswap v3 pool of tokens $X$ and $Y$ and let $p$ be the current price of token $X$ in terms of token $Y$. Suppose that a liquidity provider deposits an amount $A_0$ of token $X$ and an amount $B_0$ of token $Y$ in the price interval $[p_a, p_b]$ satisfying that $\sqrt{p_a p_b} = p$, and let $L$ be the corresponding liquidity parameter of the position. From Table 5-1, we know that

$$A_0 = L \left( \frac{1}{\sqrt{p}} - \frac{1}{\sqrt{p_b}} \right) \quad \text{and} \quad B_0 = L \left( \sqrt{p} - \sqrt{p_a} \right).$$

Let

$$r = \sqrt{\frac{p_b}{p_a}}.$$

Note that

$$p_a = \frac{p}{r} \quad \text{and} \quad p_b = rp$$

since

$$rp_a = \sqrt{\frac{p_b}{p_a}} p_a = \sqrt{p_a p_b} = p$$

and

$$rp = \sqrt{\frac{p_b}{p_a}} \sqrt{p_a p_b} = p_b.$$

Now, observe that

$$\frac{B_0}{A_0} = \frac{L\left(\sqrt{p} - \sqrt{p_a}\right)}{L\left(\frac{1}{\sqrt{p}} - \frac{1}{\sqrt{p_b}}\right)} = \frac{\sqrt{p} - \frac{\sqrt{p}}{\sqrt{r}}}{\frac{1}{\sqrt{p}} - \frac{1}{\sqrt{p}\sqrt{r}}} = \frac{\sqrt{p}\left(1 - \frac{1}{\sqrt{r}}\right)}{\frac{1}{\sqrt{p}}\left(1 - \frac{1}{\sqrt{r}}\right)} = p.$$

Hence, since the current price is $p$, the liquidity provider can deposit an amount $A_0$ of token $X$ and an amount $B_0$ of token $Y$ into a Uniswap v2 pool (this follows from Equations 2.4 and 2.12). The liquidity parameter of this Uniswap v2 position is

$$L_2 = \sqrt{A_0 B_0}.$$

Therefore, the ratio between the liquidity parameters of the Uniswap v3 and Uniswap v2 pools is

$$\frac{L}{L_2} = \frac{L}{\sqrt{A_0 B_0}} = \frac{L}{\sqrt{L\left(\frac{1}{\sqrt{p}} - \frac{1}{\sqrt{p_b}}\right)L\left(\sqrt{p} - \sqrt{p_a}\right)}}$$

$$= \frac{1}{\sqrt{\left(\frac{1}{\sqrt{p}} - \frac{1}{\sqrt{r}\sqrt{p}}\right)\left(\sqrt{p} - \frac{\sqrt{p}}{\sqrt{r}}\right)}}$$

$$= \frac{1}{\sqrt{\frac{1}{\sqrt{p}}\left(1 - \frac{1}{\sqrt{r}}\right)\sqrt{p}\left(1 - \frac{1}{\sqrt{r}}\right)}} = \frac{1}{\sqrt{\left(1 - \frac{1}{\sqrt{r}}\right)^2}} = \frac{1}{1 - \frac{1}{\sqrt{r}}}$$

$$= \frac{1}{1 - \left(\frac{p_a}{p_b}\right)^{\frac{1}{4}}}.$$

**Example 5.6.** In Table 5-5, we show the (approximate) values of the quotient $\frac{L}{L_2}$ for different values of the parameter $r$ together with the corresponding values of the quotient $\frac{p_b}{p_a}$. For instance, if $r = 1.1$, then $p_b = 1.21p_a$, which means that price $p_b$ is 21% higher that price $p_a$. In such an interval, the liquidity parameter of a Uniswap v3 position will be approximately 21.5 times higher than the liquidity parameter of the same liquidity deposit into a Uniswap v2 pool.

**Table 5-5.** *Values of the ratios between the liquidity parameters of a Uniswap v3 and a Uniswap v2 position for different values of the parameter r that measures the length of the price interval of Uniswap v3*

| $r$ | $\dfrac{p_b}{p_a}$ | $\dfrac{L}{L_2}$ |
|---|---|---|
| 1.005 | 1.010025 | 401.5 |
| 1.01 | 1.0201 | 201.5 |
| 1.05 | 1.1025 | 41.5 |
| 1.1 | 1.21 | 21.5 |
| 1.2 | 1.44 | 11.5 |
| 2 | 4 | 3.41 |
| 10 | 100 | 1.46 |
| 100 | 10,000 | 1.11 |

If $r = 1.05$, then $p_b = 1.1025p_a$, and hence, $p_b$ is approximately 10% higher than $p_a$. In this case, the liquidity parameter of a Uniswap v3 position is approximately 41.5 times higher than that of the same deposit into a Uniswap v2 pool. As expected, if the interval in which liquidity is deposited is smaller, then the liquidity parameter will be bigger since if the same amount of tokens is used for trading in a smaller interval, then we can provide more liquidity into that interval.

Of course, we have to remember that if the interval in which the liquidity provider provided liquidity is smaller, then the likelihood of the price falling out of that interval will be greater, and so will be the impermanent loss (see Figure 5-5). In addition, when the price goes out of that interval, the liquidity provider will stop earning trading fees.

## 5.7.2 Independence with Respect to Other Liquidity Providers

We will now prove that the amount of fees that a liquidity provider earns in a Uniswap v3 pool depends only on the price movement and the parameters that define their position and not on the positions of the other liquidity providers that have deposited liquidity on the same pool. We formalize this statement in the following proposition.

**Proposition 5.2.** *Consider a Uniswap v3 liquidity pool with tokens X and Y and fee $\phi$. Suppose that a liquidity provider owns a position in the pool. For each positive real number p, let $x_r(p)$ and $y_r(p)$ be the amount of real reserves at price p of tokens X and Y, respectively, of the liquidity provider's position.*

*Let $p_1$ and $p_2$ be positive real numbers. Suppose that a single trade in the pool moves the price from $p_1$ to $p_2$.*

- *If $p_1 < p_2$, then the liquidity provider earns an amount*

$$\frac{\phi}{1-\phi}\left(y_r\left(p_2\right)-y_r\left(p_1\right)\right)$$

*of token Y in fees.*

- *If $p_1 > p_2$, then the liquidity provider earns an amount*

$$\frac{\phi}{1-\phi}\left(x_r\left(p_2\right)-x_r\left(p_1\right)\right)$$

*of token X in fees.*

*Proof.* By dividing the price movement into several steps, we may assume that the total liquidity of the pool remains constant within the trade. Let $L_{\text{tot}}$ be the total liquidity of the pool corresponding to the interval in which the trade is executed. For each positive real number p, let $x_v(p)$

and $y_v(p)$ be the virtual balances of the pool at price $p$ of tokens $X$ and $Y$, respectively. Recall that

$$x_v(p) = \frac{L_{tot}}{\sqrt{p}} \quad \text{and} \quad y_v(p) = L_{tot}\sqrt{p}.$$

Let $L$ be the liquidity parameter of the liquidity provider's position, and let $[p_a, p_b]$ be the price interval of the position. For all positive real numbers $p$, $q$, let $F(p, q)$ be the amount of trading fees that the liquidity provider's position earns when the price moves from $p$ to $q$ in a single trade. We will first prove the following assertion, which will be called assertion (A).

**(A)** For all positive real numbers $p$, $q$, if $p_a \leq p < q \leq p_b$,
then $F(p,q) = \frac{\phi}{1-\phi}(y_r(q) - y_r(p))$.

Let $p$ and $q$ be positive real numbers such that $p_a \leq p < q \leq p_b$ and suppose that a single trade moves the price from $p$ to $q$. Since $p < q$, the amount of token $Y$ in the pool increases, and the amount of token $X$ in the pool decreases. Let $b$ be the amount of token $Y$ that is deposited into the pool. Hence, the trading fee will be charged on this amount $b$ of token $Y$. Note that the amount of the fee is $\phi b$, and then the amount of token $Y$ that goes into the pool is $(1 - \phi)b$. Thus, $y_v(q) = y_v(p) + (1 - \phi)b$. Therefore,

$$F(p,q) = \frac{L}{L_{tot}}\phi b = \frac{L\phi}{L_{tot}(1-\phi)}(y_v(q) - y_v(p))$$

$$= \frac{L\phi}{L_{tot}(1-\phi)}\left(L_{tot}\sqrt{q} - L_{tot}\sqrt{p}\right) = \frac{\phi}{1-\phi}\left(L\sqrt{q} - L\sqrt{p}\right)$$

$$= \frac{\phi}{1-\phi}\left(\left(L\sqrt{q} - L\sqrt{p_a}\right) - \left(L\sqrt{p} - L\sqrt{p_a}\right)\right)$$

$$= \frac{\phi}{1-\phi}(y_r(q) - y_r(p)),$$

where the last equality follows from the formulae of Table 5-1. Hence, we have proved assertion (A).

Now we will prove the following assertion, which will be called assertion (B).

(B) For all positive real numbers $p$, $q$, if $p_a \leq q < p \leq p_b$, then $F(p,q) = \dfrac{\phi}{1-\phi}\left(x_r(q) - x_r(p)\right)$.

Let $p$ and $q$ be positive real numbers such that $p_a \leq q < p \leq p_b$ and suppose that a single trade moves the price from $p$ to $q$. Since $p > q$, the amount of token $X$ in the pool has to increase. Let $a$ be the amount of token $X$ that is deposited into the pool. Hence, the amount of the trading fee is $\phi a$, and then the amount of token $X$ that goes into the pool is $(1 - \phi)a$. Thus, $x_v(q) = x_v(p) + (1 - \phi)a$. Therefore,

$$F(p,q) = \frac{L}{L_{tot}}\phi a = \frac{L\phi}{L_{tot}(1-\phi)}\left(x_v(q) - x_v(p)\right)$$

$$= \frac{L\phi}{L_{tot}(1-\phi)}\left(\frac{L_{tot}}{\sqrt{q}} - \frac{L_{tot}}{\sqrt{p}}\right) = \frac{\phi}{1-\phi}\left(\frac{L}{\sqrt{q}} - \frac{L}{\sqrt{p}}\right)$$

$$= \frac{\phi}{1-\phi}\left(\left(\frac{L}{\sqrt{q}} - \frac{L}{\sqrt{p_b}}\right) - \left(\frac{L}{\sqrt{p}} - \frac{L}{\sqrt{p_b}}\right)\right)$$

$$= \frac{\phi}{1-\phi}\left(x_r(q) - x_r(p)\right),$$

where the last equality follows from the formulae of Table 5-1. Hence, we have proved assertion (B).

We will now prove the first statement of the proposition. Suppose that $p_1 < p_2$. Since the price increases, an amount of token $Y$ is deposited, and thus, the trading fee is charged on token $Y$. We will divide our analysis into several cases and subcases.

Case 1: $p_1 \geq p_b$.

In this case, $p_b \le p_1 < p_2$, and thus, the price movement is outside the position range. Hence, $y_r(p_1) = y_r(p_2)$, and the amount $F(p_1, p_2)$ of trading fees that the position earns is 0. Observe that

$$F(p_1, p_2) = 0 = \frac{\phi}{1-\phi}\left(y_r(p_2) - y_r(p_1)\right),$$

that is, the first formula of the statement of the proposition holds.

Case 2: $p_2 \le p_a$.

In this case, $p_1 < p_2 \le p_a$, and hence, the price movement is outside the position range. Then the amount $F(p_1, p_2)$ of trading fees that the position earns is 0. Note that $y_r(p_1) = y_r(p_2) = 0$ since $p_1 < p_2 \le p_a$. Thus,

$$F(p_1, p_2) = 0 = \frac{\phi}{1-\phi}\left(y_r(p_2) - y_r(p_1)\right),$$

and hence, the first formula of the statement of the proposition holds.

Case 3: $p_1 < p_b$ and $p_2 > p_a$.

We divide the analysis of this case into four subcases.

Case 3.1: $p_1 < p_a$ and $p_a < p_2 \le p_b$.

Note that $y_r(p_1) = y_r(p_a) = 0$ since $p_1 < p_a$ and that the position does not earn fees within the interval $(p_1, p_a)$. Hence, the amount of trading fees that the position earns is

$$F(p_1, p_2) = F(p_a, p_2) = \frac{\phi}{1-\phi}\left(y_r(p_2) - y_r(p_a)\right)$$

$$= \frac{\phi}{1-\phi}\left(y_r(p_2) - y_r(p_1)\right)$$

where the second equality holds by assertion (A).

Case 3.2: $p_1 < p_a$ and $p_2 > p_b$.

Note that $y_r(p_1) = y_r(p_a) = 0$ since $p_1 < p_a$ and that $y_r(p_2) = y_r(p_b)$ since $p_2 > p_b$. Note also that the position does not earn fees within the intervals $(p_1, p_a)$ and $(p_b, p_2)$. Thus, the amount of trading fees that the position earns is

$$F(p_1, p_2) = F(p_a, p_b) = \frac{\phi}{1-\phi}(y_r(p_b) - y_r(p_a))$$
$$= \frac{\phi}{1-\phi}(y_r(p_2) - y_r(p_1)),$$

where the second equality holds by assertion (A).

Case 3.3: $p_a \le p_1 < p_b$ and $p_a < p_2 \le p_b$.

By assertion (A), the amount of trading fees that the position earns is

$$F(p_1, p_2) = \frac{\phi}{1-\phi}(y_r(p_2) - y_r(p_1)).$$

Case 3.4: $p_a \le p_1 < p_b$ and $p_2 > p_b$.

Note that $y_r(p_2) = y_r(p_b)$ since $p_2 > p_b$ and that the position does not earn fees within the interval $(p_b, p_2)$. Thus, the amount of trading fees that the position earns is

$$F(p_1, p_2) = F(p_1, p_b) = \frac{\phi}{1-\phi}(y_r(p_b) - y_r(p_1))$$
$$= \frac{\phi}{1-\phi}(y_r(p_2) - y_r(p_1)),$$

where the second equality holds by assertion (A).

Therefore, we have proved the first statement of the proposition.

Clearly, the second statement of the proposition can be proved in a similar way using assertion (B) instead of assertion (A).  □

# 5.8 Summary

In this chapter, we gave a comprehensive description of the Uniswap v3 AMM, and we thoroughly explained how it is implemented. In addition, we delved deeply into its smart contract and described the variables and methods that are used in the actual code. We also analyzed the impermanent losses of Uniswap v3 positions and compared them with those of similar Uniswap v2 positions.

# References

[1] Adams, H., Zinsmeister, N., and Robinson, D. Uniswap v2 Core. `https://uniswap.org/whitepaper.pdf`.

[2] Adams, H., Zinsmeister, N., Salem, M., Keefer, R., and Robinson, D. Uniswap v3 Core.

`https://uniswap.org/whitepaper-v3.pdf`.

[3] Aigner, A. A., and Dhaliwal, G. Uniswap: Impermanent loss and risk profile of a liquidity provider. *arXiv preprint arXiv:2106.14404* (2021).

`https://arxiv.org/abs/2106.14404`.

[4] Angeris, G., Evans, A., and Chitra, T. Replicating market makers. *arXiv preprint arXiv:2103.14769* (2021).

`https://arxiv.org/abs/2103.14769`.

[5] Angeris, G., Evans, A., and Chitra, T. Replicating Monotonic Payoffs Without Oracles. *arXiv preprint arXiv:2111.13740* (2021).

`https://arxiv.org/abs/2111.13740`.

[6] Angeris, G., Kao, H.-T., Chiang, R., Noyes, C., and Chitra, T. An analysis of Uniswap markets. *arXiv preprint arXiv:1911.03380* (2019).

`https://arxiv.org/abs/1911.03380`.

REFERENCES

[7]   Antonopoulos, A. M., and Wood, G. *Mastering Ethereum: Building Smart Contracts and DApps*. O'Reilly Media, 2018.

[8]   Berg, J. A., Fritsch, R., Heimbach, L., and Wattenhofer, R. An Empirical Study of Market Inefficiencies in Uniswap and SushiSwap. *arXiv preprint arXiv:2203.07774* (2022).

      https://arxiv.org/abs/2203.07774.

[9]   Buterin, V. Let's run on-chain decentralized exchanges the way we run prediction markets. *Reddit post* (2016).

      www.reddit.com/r/ethereum/comments/55m04x/
      lets_run_onchain_decentralized_exchanges_
      the_way/.

[10]  Cvetkovski, Z. *Inequalities: Theorems, Techniques and Selected Problems*. SpringerLink : Bücher. Springer Berlin Heidelberg, 2012.

[11]  Egorov, M. StableSwap - efficient mechanism for Stablecoin liquidity.

      https://curve.fi/files/stableswap-paper.pdf.

[12]  Egorov, M. Automatic market-making with dynamic peg.

      https://curve.fi/files/crypto-pools-
      paper.pdf.

[13]  Elsts, A. Liquidity math in Uniswap v3.

      https://atiselsts.github.io/pdfs/uniswap-v3-
      liquidity-math.pdf.

[14]  Engel, D., and Herlihy, M. Composing networks
      of automated market makers. In *Proceedings of
      the 3rd ACM Conference on Advances in Financial
      Technologies* (2021), pp. 15–28.

      https://arxiv.org/abs/2106.00083.

[15]  Evans, A. Liquidity provider returns in
      geometric mean markets. *arXiv preprint
      arXiv:2006.08806* (2020).

      https://arxiv.org/abs/2006.08806.

[16]  Fritsch, R. Concentrated Liquidity in Automated
      Market Makers. In *Proceedings of the 2021 ACM CCS
      Workshop on Decentralized Finance and Security*
      (2021), pp. 15–20.

      https://arxiv.org/abs/2110.01368.

[17]  Group, A. Automated Market Makers (AMMs):
      Versioning Up. *Medium post* (2022).

      https://ambergroup.medium.com/automated-
      market-makers-amms-versioning-up-
      2c8a81e9889f.

[18]  Hanson, R. Combinatorial information market
      design. *Information Systems Frontiers 5*, 1 (2003),
      107–119.

[19]  Heimbach, L., and Wattenhofer, R. Eliminating
      Sandwich Attacks with the Help of Game Theory.
      *arXiv preprint arXiv:2202.03762* (2022).

      https://arxiv.org/abs/2202.03762.

[20]   Heimbach, L., and Wattenhofer, R. SoK: Preventing Transaction Reordering Manipulations in Decentralized Finance. *arXiv preprint arXiv:2203.11520* (2022).

https://arxiv.org/abs/2203.11520.

[21]   Hertzog, E., Benartzi, G., and Benartzi, G. Bancor Protocol. Continuous Liquidity for Cryptographic Tokens Through Their Smart Contracts.

https://cryptorating.eu/whitepapers/Bancor/bancor_protocol_whitepaper_en.pdf.

[22]   Jensen, J. R., Pourpouneh, M., Nielsen, K., and Ross, O. The Homogenous Properties of Automated Market Makers. *arXiv preprint arXiv:2105.02782* (2021).

https://arxiv.org/abs/2105.02782.

[23]   Krishnamachari, B., Feng, Q., and Grippo, E. Dynamic curves for decentralized autonomous cryptocurrency exchanges. *arXiv preprint arXiv:2101.02778* (2021).

https://arxiv.org/abs/2101.02778.

[24]   Larson, R., and Edwards, B. H. *Calculus of a single variable*. Cengage Learning, 2016.

[25]   Loesch, S., Hindman, N., Richardson, M. B., and Welch, N. Impermanent Loss in Uniswap v3. *arXiv preprint arXiv:2111.09192* (2021).

https://arxiv.org/abs/2111.09192.

[26] Manfrino, R. B., Gómez Ortega, J. A., and Valdez Delgado, R. *Inequalities: A Mathematical Olympiad Approach*. Birkhäuser Basel, 2009.

[27] Martinelli, F., and Mushegian, N. Balancer whitepaper - A non-custodial portfolio manager, liquidity provider, and price sensor.

    https://balancer.fi/whitepaper.pdf.

[28] McMenamin, C., Daza, V., and Fitzi, M. FairTraDEX: A Decentralised Exchange Preventing Value Extraction. *arXiv preprint arXiv:2202.06384* (2022).

    https://arxiv.org/abs/2202.06384.

[29] Mohan, V. Automated market makers and decentralized exchanges: a Defi primer. *Financial Innovation 8*, 1 (2022), pp. 1–48.

[30] Nguyen, A., and Luu, L. Dynamic Automated Market Making.

    https://files.kyber.network/DMM-Feb21.pdf.

[31] Niemerg, A., Robinson, D., and Livnev, L. YieldSpace: An Automated Liquidity Provider for Fixed Yield Tokens.

    https://yield.is/YieldSpace.pdf.

[32] Pourpouneh, M., Nielsen, K., and Ross, O. Automated Market Makers.

[33]  semaji.eth. In the Long Run, We Are All Dead: The Fate of Passive Retail Liquidity Provision in Uniswap v3.

```
https://semaji.substack.com/p/in-the-
long-run-we-are-all-dead-the?utm_
source=twitter&s=r.
```

[34]  Tiruviluamala, N., Port, A., and Lewis, E. A General Framework for Impermanent Loss in Automated Market Makers. arXiv preprint arXiv:2203.11352 (2022).

```
https://arxiv.org/abs/2203.11352.
```

[35]  Wang, Y. Automated Market Makers for Decentralized Finance (DeFi). *arXiv preprint arXiv:2009.01676* (2020).

```
https://arxiv.org/abs/2009.01676.
```

[36]  Wang, Y. Implementing Automated Market Makers with Constant Circle. *arXiv preprint arXiv:2103.03699* (2021).

```
https://arxiv.org/abs/2103.03699.
```

[37]  Wood, G. Ethereum: A secure decentralised generalised transaction ledger. *Ethereum project yellow paper* (2014).

[38]  Wu, M., and McTighe, W. Constant Power Root Market Makers. *arXiv preprint arXiv:2205.07452* (2022).

```
https://arxiv.org/abs/2205.07452.
```

[39]   Xu, J., Paruch, K., Cousaert, S., and Feng, Y. Sok: Decentralized Exchanges (DEX) with Automated Market Maker (AMM) protocols. *arXiv preprint arXiv:2103.12732* (2021).

https://arxiv.org/abs/2103.12732.

[40]   Zamyatin, A., Wolter, K., Werner, S., Harrison, G., Mulligan, C. E., and Knottenbelt, W. J. Swimming with fishes and sharks: Beneath the surface of queue-based ethereum mining pools. In *2017 IEEE 25th International Symposium on Modeling, Analysis, and Simulation of Computer and Telecommunication Systems (MASCOTS)* (2017), IEEE, pp. 99–109.

[41]   Zhang, A. L. Nested AMMs.

https://github.com/anthonyleezhang/nestedamm/blob/main/nestedamm.pdf.

[42]   Zhou, L., Qin, K., and Gervais, A. A2MM: Mitigating frontrunning, transaction reordering and consensus instability in decentralized exchanges. *arXiv preprint arXiv:2106.07371* (2021).

https://arxiv.org/abs/2106.07371

# Index

© Miguel Ottina, Peter Johannes Steffensen, Jesper Kristensen 2023
M. Ottina et al., *Automated Market Makers*, https://doi.org/10.1007/978-1-4842-8616-6

# N, O, P, Q, R

# S, T

# U, V, W, X, Y, Z

Printed in the United States
by Baker & Taylor Publisher Services